Approximation Techniques for Engineers

SECOND EDITION

Approximation Techniques for Engineers

SECOND EDITION

Dr. Louis Komzsik

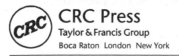

CRC Press
Taylor & Francis Group
Boca Raton London New York

CRC Press is an imprint of the
Taylor & Francis Group, an **informa** business

CRC Press
Taylor & Francis Group
6000 Broken Sound Parkway NW, Suite 300
Boca Raton, FL 33487-2742

First issued in paperback 2020

ISBN-13: 978-1-138-70005-5 (hbk)
ISBN-13: 978-0-367-65807-6 (pbk)

Library of Congress Cataloging-in-Publication Data

Names: Komzsik, Louis, author.
Title: Approximation techniques for engineers / Louis Komzsik.
Description: Second edition. | Boca Raton : Taylor & Francis, CRC Press, 2017. | Includes bibliographical references.
Identifiers: LCCN 2017010757| ISBN 9781138700055 (hardback : alk. paper) | ISBN 9781315205007 (ebook)
Subjects: LCSH: Approximation theory. | Engineering--Statistical methods.
Classification: LCC QA221 .K665 2017 | DDC 511/.4--dc23
LC record available at https://lccn.loc.gov/2017010757

Visit the Taylor & Francis Web site at
http://www.taylorandfrancis.com

and the CRC Press Web site at
http://www.crcpress.com

Although this may seem as a paradox, all exact science
is dominated by the idea of approximation

Bertrand Russell

Contents

Preface to the second edition

Since the original publication of this book a decade ago, I have received several comments and recommendations from readers in academia and industry. They pointed out certain topics that fit into existing chapters, resulting in several new sections and subsections. They also requested two entirely new chapters as an extension. This edition is based on those recommendations.

The added sections address the approximation of matrix functions (4.6 and 4.6.1), techniques of deflating the eigenvalue solution space (10.1.1 and 10.1.2) and improving the accuracy of approximate eigenvalue solutions (10.1.3 and 10.1.4). Furthermore, the successive approximation solution of initial value problems of systems of differential equations (11.4.1 and 11.4.3), the linearization of second order initial value problems (11.6), the model order reduction of the transient response problem (11.7.1) and the approximate solution of boundary value problems (12.1.3) are also presented.

The new Chapter 13 starts with the conversion of initial and boundary value problems to integral equations. The topic of integral equations is seldom covered in numerical texts and there are no general purpose software tools to solve such problems. Hence algorithms to solve the main classes of these problems and illustrative hand examples are presented in this new chapter.

Optimization, the topic of Chapter 14, while the subject of many specialized books, is mainly described with direct application solutions and less as a mathematical approximation. This chapter explains the fundamental concepts and the mainstream methods, again with examples. Both of these new chapters conclude with an application section.

The handful of typos and errors found in the first edition during the past decade were corrected in this edition. It is hoped that the new content is as error free as possible, but I am solely responsible for any remaining errors.

Louis Komzsik

2016

Preface

This book is a collection of methods that provide an approximate result for certain engineering computations. The difference from the analytical result (if such exists at all) occurs due to the formulation chosen to execute a computation. As such, it must be distinguished from numerical errors occurring due to the computational round-off error of the finite precision of computers.

There is a perennial quest by mathematicians and engineers to find approximate results in two classes of problems. In one class, the input data may only be given by a discrete set of points to describe the continuous geometry of a physical phenomenon; however, a continuous function fitting the data is sought. It is also possible in this class that the input data is given by functions that need to be approximated by another function or some discrete quantities of the functions (such as the derivative at certain point) are required. In the other class, usually a continuous problem is posed and an approximate solution at a discrete set of points is desired, for practical reasons such as computational cost.

Accordingly, the book is divided into two parts: data approximation techniques and approximate solutions. The first part starts with the classical interpolation methods, followed by spline interpolations and least square approximations. It also covers various approximations of functions as well as their numerical differentiation and integration. The second part ranges from the solution of algebraic equations, linear and nonlinear systems, through eigenvalue problems to initial and boundary value problems. Both parts emphasize the logical thread and common principles of the approximation techniques.

The book is intended to be an everyday tool as a reference for practicing engineers, researchers and graduate engineering students. It is hoped that readers can solve a particular approximation problem arising in their practice by directly focusing on a chapter or section describing the appropriate techniques.

Louis Komzsik

2006

Acknowledgments

Dr. Amy Sohn, Siemens PLM Software, deserves my appreciation for her diligent proofreading and corrections of the second edition. My thanks are due to Dr. Leonard Hoffnung, MathWorks Inc., one of the reviewers of the first edition, for also reviewing the additions to this version and providing valuable recommendations.

Dr. Thomas Flöck of Siemens Digital Factory meticulously reviewed the new edition with the mindset of the control system analyst. Dr. Daniel Heiserer of BMW Electric Car Development represented the automotive engineer's focus and provided the image in the cover art depicting the various stages of approximations during the development of the i8, ranging from sketches through finite element models to the final geometry.

Finally, I am thankful to Publisher Nora Konopka, Engineering Editor Michele Smith, Production Editor Michele Dimont, Editorial Assistant Kyra Lindholm, and Cover Artist John Gandour at Taylor & Francis, for their dedicated support during the publication process, hopefully making this second edition a success.

About the author

Dr. Komzsik is a graduate of the Technical University and the Eötvös University of Sciences, both in Budapest, Hungary and worked for the Hungarian Shipyards in Budapest during the 1970s as an engineering analyst. After immigrating to the United States in 1981, he worked for the McDonnell-Douglas Corporation as a senior analyst, at the MacNeal-Schwendler (now MSC Software) Corporation as Chief Numerical Analyst and he recently retired as Principal Key CAE Expert from Siemens PLM Software.

He is the author of the original *NASTRAN Numerical Methods Handbook*, and a widely read book on *The Lanczos Method* that was also published in Chinese, Japanese and Hungarian. His books titled *Computational Techniques of Finite Element Analysis*, *Applied Calculus of Variations for Engineers* and *Approximation Techniques for Engineers* are already in their second editions. He is also the co-author of the book *Rotor Dynamics with Finite Elements*.

Part I

Data approximations

1

Classical interpolation methods

The first approximation technique discussed here, and one of the earliest of such, is interpolation. The word was coined by Wallis in 1656 [8] and there were already earlier attempts at similar techniques. An interpolation method, nowadays called Newton's interpolation, appeared in Newton's classical book in 1687 [4], and is the subject of the first section.

Lagrange, the author of the topic of the second section, gave acknowledgment to Newton when publishing his method in 1795 [3]. Hermite [1] generalized the interpolation problem in 1878 to also consider the derivatives of the approximated function. Finally, approximation of functions of two variables was first considered by Picard in 1891 [5].

These methods, as their dates indicate, are rather old and fundamental. While they originated earlier, they were later supported by Weierstrass' approximation theorem [9] dated from 1885. The theorem states that for a function $f(x)$ that is continuous in the interval $[a, b]$ there exists a polynomial $p(x)$ such that

$$|f(x) - p(x)| < \epsilon$$

for $a \leq x \leq b$, where ϵ is an arbitrary small value. There are several ways to find such polynomials, the most important being the Lagrange, Newton and Hermite methods discussed in this chapter.

An important aspect of these techniques is that the function $f(x)$ is sampled at a certain number of points. These points $(x_i, f(x_i)), i = 0, \ldots n$, are chosen to find the approximating polynomial. This approach also occurs naturally, when the $f(x)$ function is not known explicitly, and only a set of discrete points are given; hence these methods belong to the data approximation class.

Their practical use as a tool for interpolating a given set of points is somewhat less important than their application as a building component of more advanced techniques, as will be shown in later chapters.

1.1   Newton interpolation

Newton's method approximates a function $f(x)$ given by a set of n points

$$(x_k, f(x_k)), k = 0, 1, 2, \ldots n,$$

with a polynomial of the form

$$p_N(x) = a_0 + a_1(x - x_0) + a_2(x - x_0)(x - x_1) + \ldots + a_n(x - x_0) \cdots (x - x_{n-1}).$$

The a_k coefficients of Newton's approximation may be computed from the condition of the polynomial going through all given points. Hence

$$f(x_0) = a_0,$$

and

$$f(x_1) = f(x_0) + a_1(x_1 - x_0),$$

from which

$$a_1 = \frac{f(x_1) - f(x_0)}{x_1 - x_0}$$

follows. Let us introduce the so-called divided difference notation. The first order divided differences are of the form

$$f[x_k, x_{k+1}] = \frac{f(x_{k+1}) - f(x_k)}{x_{k+1} - x_k}.$$

In this notation

$$a_1 = f[x_0, x_1].$$

The notation for second order differences is

$$f[x_k, x_{k+1}, x_{k+2}] = \frac{f[x_{k+1}, x_{k+2}] - f[x_k, x_{k+1}]}{x_{k+2} - x_k}.$$

Generalizing the notation to the *i*th differences gives the Newton interpolation polynomial as

$$p_N(x) = f[x_0] + f[x_0, x_1](x - x_0) + f[x_0, x_1, x_2](x - x_0)(x - x_1) + \ldots,$$

with $f[x_0] = f(x_0)$. Introducing $\omega_0 = 1$ and the polynomial

$$\omega_k = (x - x_0)(x - x_1) \cdots (x - x_{k-1})$$

yields the most compact (and easy to memorize) form of Newton's interpolation polynomial:

$$p_N(x) = \sum_{k=0}^{n} f[x_0, \ldots, x_k] \omega_k.$$

The error of the approximation of this method is

$$f(x) - p_N(x) = f[x_0, x_1, \ldots, x_n, x] \omega_n$$

where

$$f[x_0, x_1, \ldots, x_n, x] = \frac{f^{(n+1)}(\xi)}{(n+1)!}$$

for some $x_0 \leq \xi \leq x_n$.

1.1.1 Equidistant Newton interpolation

Naturally, the case of equidistant abscissa values is of practical importance for engineers and may be exploited here. In this case an arbitrary x value may be defined as

$$x = x_0 + sh,$$

where s is a real number and h is a step size, also real. Then the given abscissa values are computed as

$$x_k = x_0 + kh,$$

where k is an integer counter. With this, Newton's interpolation polynomial becomes

$$p_N(x) = p_N(x_0 + sh) = f[x_0] + f[x_0, x_1]sh + f[x_0, x_1, x_2]s(s-1)h^2 + \ldots,$$

which, after some algebraic manipulations, results in

$$p_N(x) = f[x_0] + \sum_{k=1}^{n} \binom{s}{k} k! h^k f[x_0, \ldots, x_k].$$

The latter is called Newton's forward divided difference formula because the starting point was the leftmost point of the given set and we propagated to the right. Therefore this formula has more accurate approximation in the neighborhood of the leftmost point.

Since the points are equidistant, the divided differences may be simplified by simple differences as follows:

$$f[x_0, x_1] = \frac{f(x_1) - f(x_0)}{x_1 - x_0} = \frac{1}{h} \Delta f(x_0),$$

$$f[x_0, x_1, x_2] = \frac{\Delta f(x_1) - \Delta f(x_0)}{h} = \frac{1}{2h^2} \Delta^2 f(x_0),$$

and so on for more equidistant points. With this notation, we obtain the class of Newton's forward difference formula (note the omission of the word divided) as follows:

$$p_N(x) = f(x_0) + \sum_{k=1}^{n} \binom{s}{k} \Delta^k f(x_0).$$

It is also possible to start from the rightmost point and propagate backward. Newton's backward divided difference formula may be written as

$$p_N^B(x) = f[x_n] + f[x_n, x_{n-1}](x - x_n) + f[x_n, x_{n-1}, x_{n-2}](x - x_n)(x - x_{n-1}) + \dots.$$

The equidistant case for the backward divided difference formula follows the derivation above:

$$p_N^B(x) = f[x_n] + f[x_n, x_{n-1}]sh + f[x_n, x_{n-1}, x_{n-2}]s(s+1)h^2 + \dots.$$

Introducing now the backward differences of

$$\nabla f(x_i) = f(x_{i-1}) - f(x_i),$$

Newton's backward difference formula is written as

$$p_N^B(x) = f(x_0) + \sum_{k=1}^{n} (-1)^k \binom{-s}{k} \nabla^k f(x_n).$$

The above formulae are advantageous in the neighborhood of the rightmost point of the given set.

Finally, when the best accuracy is required in the middle of the approximation interval, a so-called centered difference formula, also known as Stirling's formula, may be used. In the centered difference approach we denote the point closest to the point of interpolation x as x_0 and employ the following indexing scheme:

$$x \approx x_0 < x_1 < x_2 < \dots < x_m,$$

and

$$x_{-m} < x_{-m+1} < \dots < x_{-2} < x_{-1} < x \approx x_0.$$

This implies the presence of an odd number $n = 2m + 1$ of points. In case there are an even number of points given, the left-hand sequence ends in index $-m+1$. Focusing on the equidistant case, with five given points, the formula is

$$p_N^C(x) = f[x_0] + \frac{1}{2}s \cdot h(f[x_{-1}, x_0] + f[x_0, x_1]) + s^2 h^2 f[x_{-1}, x_0, x_1]$$

$$+ \frac{1}{2}s(s^2 - 1)h^3(f[x_{-2}, x_{-1}, x_0, x_1] + f[x_{-1}, x_0, x_1, x_2]).$$

If $n = 4$, the same formula may be used without the first half of the last term, as the numbering strategy for four points indicates:

$$x_{-1}, x_0, x_1, x_2.$$

The formula may be extended for a higher number of points: for a given m value the last two generic terms are

$$s^2(s^2 - 1)(s^2 - 4) \cdots (s^2 - (m - 1)^2)h^{2m} f[x_{-m}, \ldots, x_m]$$

$$+ \frac{1}{2}s(s^2 - 1) \cdots (s^2 - m^2)h^{2m+1}(f[x_{-m-1}, \ldots, x_m] + f[x_{-m}, \ldots, x_{m+1}a]).$$

If n is even, then the first part of the last term is again omitted.

There are also other families of interpolation strategies based on divided differences, such as the Gauss, Bessel and Everett formulae. These are well detailed and even tabulated in some references, for example [2].

1.1.2 Computational example

The following computational example demonstrates some of the important fundamental aspects and the actual computational mechanism. Let us consider the following set of points:

$$(x, y) = (4, 1); (6, 3); (8, 8); (10, 20).$$

The set contains four points that will result in a third order interpolation polynomial.

We use Newton's divided difference formula for the problem. The first order divided differences are

$$f[x_0, x_1] = \frac{3 - 1}{6 - 4} = 1,$$

$$f[x_1, x_2] = \frac{8 - 3}{8 - 6} = 5/2,$$

and

$$f[x_2, x_3] = \frac{20 - 8}{10 - 8} = 6.$$

Based on these, the second order differences are

$$f[x_0, x_1, x_2] = \frac{5/2 - 1}{8 - 4} = 3/8,$$

and
$$f[x_1, x_2, x_3] = \frac{6 - 5/2}{10 - 6} = 7/8.$$
Finally, the third order divided difference is
$$f[x_0, x_1, x_2, x_3] = \frac{7/8 - 3/8}{10 - 4} = 1/12.$$
The Newton interpolation polynomial is written as
$$p_N(x) = 1 + 1(x - 4) + \frac{3}{8}(x - 4)(x - 6) + \frac{1}{12}(x - 4)(x - 6)(x - 8),$$
which simplifies to
$$p_N(x) = \frac{1}{24}(2x^3 - 27x^2 + 142x - 240).$$
Figure 1.1 shows how smooth the Newton interpolation polynomial is for these points. One can see that Newton's method satisfies the approximation condition of going through all the input points and, partially due to the smooth input data, provides a nice approximating polynomial.

Let us now consider approximating the unknown function value at various mid-segment locations of $x = 5, 7, 9$ and use the most appropriate equidistant divided difference formula with $h = 2$, but only to the second order differences. To facilitate these computations, we gather the divided differences into Table 1.1.

TABLE 1.1
Example of divided differences

i	x_i	$f(x_i)$	$f[x_i, x_{i+1}]$	$f[x_i, x_{i+1}, x_{i+2},]$	$f[x_i, x_{i+1}, x_{i+2}, x_{i+3}]$
0	4	1	-	-	-
1	6	3	1	-	-
2	8	8	5/2	3/8	-
3	10	20	6	7/8	1/12

The forward divided difference formula with $s = \frac{1}{2}$ yields
$$p_N(5) = f(5) = 1 + 1\frac{1}{2}2 + \frac{3}{8}\frac{1}{2}(\frac{1}{2} - 1)2^2 = \frac{13}{8} = 1.625.$$
Note that the forward difference formula uses the top term of each column of Table 1.1.

The backward divided difference formula with $s = -\frac{1}{2}$ yields
$$p_N^B(9) = f(9) = 20 + 6\frac{-1}{2}2 + \frac{7}{8}\frac{-1}{2}(\frac{-1}{2} + 1)2^2 = 13 + 1/8 = 13.125.$$

FIGURE 1.1 Newton interpolation polynomial

Note that the backward difference formula uses the bottom term of each column of Table 1.1.

The centered divided difference formula with $s = \frac{1}{2}$ yields

$$p_N^C(7) = f(7) = 3 + \frac{1}{2} \cdot \frac{1}{2} 2(1 + \frac{5}{2}) + \frac{1}{4} 2^2 \frac{3}{8} = 5 + 1/8 = 5.125.$$

Note that the centered difference formula uses an intermediate term of each column of Table 1.1 and the indexing is according to the rightmost column. Also, in the application of the centered difference formula we considered the fact that an even number of points (four) was given.

The results of these computations, shown in Figure 1.1 as "diff_points", are separated from the solution curve somewhat, as not all the available divided differences were used. The corresponding curve locations at $x = 5, 7, 9$ are $1.875, 4.875$ and 12.875 and the differences are proportional to the omitted third order divided difference term.

1.2 Lagrange interpolation

The Lagrange method of interpolation finds an interpolating polynomial with specifically constructed Lagrange base polynomials. They have the following characteristic:

$$L_k(x_i) = \begin{cases} 0, k \neq i, \\ 1, k = i. \end{cases}$$

With these, the Lagrange interpolation polynomial is formed as

$$p_L(x) = \sum_{k=0}^{n} f(x_k) L_k(x).$$

The base polynomials are constructed in the form of rational expressions

$$L_k = \prod_{i=0, i \neq k}^{n} \frac{x - x_i}{x_k - x_i}.$$

In an expanded form, these are

$$L_k(x) = \frac{(x - x_0) \cdots (x - x_{k-1})(x - x_{k+1}) \cdots (x - x_n)}{(x_k - x_0) \cdots (x_k - x_{k-1})(x_k - x_{k+1}) \cdots (x_k - x_n)}.$$

It follows from the above that the Lagrange interpolation polynomial satisfies the

$$p_L(x_k) = f(x_k)$$

condition, i.e., it matches the given set of points and interpolates between. The error of interpolation between the function $f(x)$ (if such was given) and the Lagrange polynomial is measured as

$$f(x) - p_L(x) = \frac{f^{(n+1)}(\xi)}{(n+1)!} \prod_{i=0}^{n} (x - x_i),$$

where $a \leq \xi \leq b$ is a location of the $(n+1)$th derivative. The proof of this formula is beyond the computational and reference focus of this book. Theoretical numerical analysis texts such as [10] present this proof.

1.2.1 Equidistant Lagrange interpolation

The general Lagrange technique is cumbersome, especially considering that one needs to recompute the L_k base polynomial when adding additional data points. The process is easier when an equidistant set of points is given or sampled. Let us describe such a set by

$$x_k - x_{k-1} = h,$$

or

$$x_k = x_0 + kh.$$

Let us further assume that

$$x = x_0 + sh,$$

where s is now not necessarily an integer, i.e., it is not a counter like k. Substituting these into the Lagrange base polynomials yields

$$L_k(x) = \frac{sh(s-1)h\cdots(s-(k-1))h(s-(k+1))h\cdots(s-n)h}{kh(k-1)h\cdots h(-h)(-2h)\cdots(-(n-k)h)}.$$

Collecting simplifies this to

$$L_k(x) = \frac{h^n s(s-1)\cdots(s-(k-1))(s-(k+1))\cdots(s-n)}{h^n k(k-1)\cdots 2\cdot 1\cdot(-1)\cdot(-2)\cdots(k-n)}$$

or

$$L_k(x) = \prod_{i=0,i\neq k}^{n} \frac{s-i}{k-i} = \ell_k(s).$$

The latter expression is independent of the h equidistant step size. It is dependent on the s value specifying the location of x and it is commonly referred to by the specialized $\ell_k(s)$ notation. As such, it may be tabulated [7] and it was used in this form before modern computers came along.

1.2.2   Computational example

Let us now approximate the set of points

$$(x, y) = (4, 1); (6, 3); (8, 8); (10, 20)$$

with Lagrange's method. The base polynomials are

$$L_0(x) = \frac{(x-6)(x-8)(x-10)}{(4-6)(4-8)(4-10)} = -\frac{1}{48}(x^3 - 24x^2 + 188x - 480),$$

$$L_1(x) = \frac{(x-4)(x-8)(x-10)}{(6-4)(6-8)(6-10)} = \frac{1}{16}(x^3 - 22x^2 + 152x - 320),$$

$$L_2(x) = \frac{(x-4)(x-6)(x-10)}{(8-4)(8-6)(8-10)} = -\frac{1}{16}(x^3 - 20x^2 + 124x - 240),$$

and

$$L_3(x) = \frac{(x-4)(x-6)(x-8)}{(10-4)(10-6)(10-8)} = \frac{1}{48}(x^3 - 18x^2 + 104x - 192).$$

Figure 1.2 shows the Lagrange base polynomials for these points along with the zero and unit value locations. Note that each takes up a value of unity at the point corresponding to its index and zero at all other input point locations.

The Lagrange interpolation polynomial approximating the given set of points is

$$p_L(x) = \sum_{k=0}^{3} f(x_k)L_k(x) = 1 \cdot L_0(x) + 3 \cdot L_1(x) + 8 \cdot L_2(x) + 20 \cdot L_3(x)$$

$$= \frac{1}{24}(2x^3 - 27x^2 + 142x - 240).$$

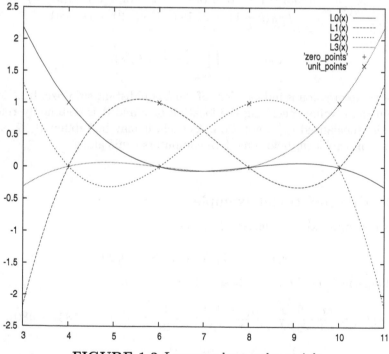

FIGURE 1.2 Lagrange base polynomials

This is identical to the result obtained by Newton's method; hence Figure 1.1 shows the Lagrange interpolation polynomial for these points as well. The value of the interpolation polynomial at an interior point, say, $x = 5$ is

$$p_L(5) = \frac{1}{24}(2 \cdot 125 - 27 \cdot 25 + 142 \cdot 5 - 240) = \frac{45}{24} = \frac{15}{8}.$$

Incidentally the example point set is equidistant with $h = 2$. Let us now see how the simplified formulation works. The value of $s = 1/2$ corresponds to the $x = 5$ location according to the above definition. Then

$$\ell_k(1/2) = \prod_{i=0, i \neq k}^{n} \frac{1/2 - i}{k - i}.$$

Specifically

$$\ell_0 = \frac{(1/2 - 1)(1/2 - 2)(1/2 - 3)}{(0 - 1)(0 - 2)(0 - 3)} = \frac{15}{48},$$

$$\ell_1 = \frac{(1/2 - 0)(1/2 - 2)(1/2 - 3)}{(1 - 0)(1 - 2)(1 - 3)} = \frac{15}{16},$$

$$\ell_2 = \frac{(1/2 - 0)(1/2 - 1)(1/2 - 3)}{(2 - 0)(2 - 1)(2 - 3)} = -\frac{5}{16},$$

and

$$\ell_3 = \frac{(1/2 - 0)(1/2 - 1)(1/2 - 2)}{(3 - 0)(3 - 1)(3 - 2)} = \frac{3}{48}.$$

The approximated value is

$$p_L(s = 1/2) = 1 \cdot \frac{15}{48} + 3 \cdot \frac{15}{16} - 8 \cdot \frac{5}{16} + 20 \cdot \frac{3}{48} = \frac{45}{24} = \frac{15}{8}.$$

This is the same value as calculated from the generic formulation, as expected. Both of the methods may be efficiently computed by hand; however, they are more instrumental in their upcoming applications in later chapters.

1.2.3 Parametric Lagrange interpolation

Parametric curves are also commonplace in engineering practice, hence a parametric rendering of Lagrange's method is also useful. Let us consider a set of points

$$Q_i = (x_i, y_i), i = 0, 1, 2, 3$$

for this computation. We can use a parametric Lagrange polynomial approximation with these four points and L_i Lagrange base polynomials as

$$p_L(x(t), y(t)) = \sum_{i=0}^{3} Q_i L_i(t).$$

If we assume that the Q_i points belong to parameter values $t = 0, 1/3, 2/3, 1$, then the following Lagrange base polynomials are used:

$$L_0(t) = \frac{(t - 1/3)(t - 2/3)(t - 1)}{(-1/3)(-2/3)(-1)} = 1 - \frac{11}{2}t + 9t^2 - \frac{9}{2}t^3,$$

$$L_1(t) = \frac{t(t - 2/3)(t - 1)}{(1/3)(-1/3)(-2/3)} = 9t - \frac{45}{2}t^2 + \frac{27}{2}t^3,$$

$$L_2(t) = \frac{t(t - 1/3)(t - 1)}{(2/3)(1/3)(-1/3)} = -\frac{9}{2}t + 18t^2 - \frac{27}{2}t^3,$$

and

$$L_3(t) = \frac{t(t - 1/3)(t - 2/3)}{(1)(2/3)(1/3)} = t - \frac{9}{2}t^2 + \frac{9}{2}t^3.$$

This may be written in a matrix form as

$$p_L(t) = \begin{bmatrix} x(t) \ y(t) \end{bmatrix} = TM_L Q,$$

where the M_L matrix is

$$M_L = \begin{bmatrix} 1 & 0 & 0 & 0 \\ -11/2 & 9 & -9/2 & 1 \\ 9 & -45/2 & 18 & -9/2 \\ -9/2 & 27/2 & -27/2 & 9/2 \end{bmatrix},$$

$$Q = \begin{bmatrix} x_0 & y_0 \\ x_1 & y_1 \\ x_2 & y_2 \\ x_3 & y_3 \end{bmatrix},$$

and

$$T = \begin{bmatrix} 1 & t & t^2 & t^3 \end{bmatrix}.$$

The final result of the parametric Lagrange approximation is

$$p_L(t) = Lx(t)\underline{i} + Ly(t)\underline{j}.$$

Applying that to the points used in the computational example section, the parametric result is

$$Lx(t) = 4 + 6t,$$

and

$$Ly(t) = 1 + 5.5t - 4.5t^2 + 18t^3.$$

This agrees with the explicit solution obtained earlier and can be seen in Figure 1.3.

FIGURE 1.3 Results of parametric Lagrange example

1.3   Hermite interpolation

Hermite interpolation generalizes the problem by also considering the derivatives of the function to be approximated. We focus on the most practical case of including the first derivative in the approximation [6]. Then the conditions of

$$f(x_k) = p_H(x_k)$$

and

$$f'(x_k) = p'_H(x_k)$$

need to be satisfied for all points $k = 0, \ldots n$. Such a polynomial may be constructed by specifically formulated Hermite base polynomials H_k.

$$p_H(x) = \sum_{k=0}^{n} f(x_k) H_k(x) + \sum_{k=0}^{n} f'(x_k) \overline{H}_k(x).$$

As there are $2n+2$ conditions to satisfy, the order of the Hermite interpolation polynomial will be $2n + 1$. There are several ways to create the Hermite base polynomials. A formulation from the Lagrange base polynomials is shown here.

$$H_k(x) = (1 - 2(x - x_k) L'_k(x_k)) L_k^2(x).$$
$$\overline{H}_k(x) = (x - x_k) L_k^2(x).$$

In the above formulae the first derivative and the square of the Lagrange base functions are used. Based on the definition of the Lagrange base functions, the Hermite base functions satisfy specific relations. When $i \neq k$, then

$$H_k(x_i) = 0,$$

and

$$\overline{H}_k(x_i) = 0.$$

When $i = k$, then

$$H_k(x_k) = (1 - 2(x_k - x_k) L'_k(x_k)) L_k^2(x_k) = (1 + 0)1 = 1,$$

and

$$\overline{H}_k = (x_k - x_k) L_k^2(x_k) = 0 \cdot 1^2 = 0.$$

From these, it follows that

$$p_H(x_k) = \sum_{i=0, i \neq k}^{n} f(x_i) \cdot 0 + f(x_k) \cdot 1 + \sum_{k=0}^{n} f'(x_k) \cdot 0 = f(x_k),$$

which is the proof of satisfying the first condition. In order to prove the satisfaction of the derivative condition, the derivative of the Hermite base functions is needed.

$$H'_k(x_i) = -2L'_k(x_k) \cdot L_k^2(x_i) + (1 - 2(x_i - x_k) L'_k(x_k)) 2 L_k(x_i) L'_k(x_i).$$

This expression is zero for both $i = k$ and $i \neq k$ for different algebraic reasons. The derivative of the $\overline{H}$ base polynomial is

$$\overline{H}'_k(x_i) = L_k^2(x_i) + (x_i - x_k) \cdot 2 L_k(x_i) L'_k(x_i).$$

This is zero when $i \neq k$ but one when $i = k$. Utilizing all the above, we can write

$$p'_H(x_k) = \sum_{k=0}^{n} f(x_k) \cdot 0 + \sum_{i=0, i \neq k}^{n} f'(x_i) \cdot 0 + f'(x_k) \cdot 1 = f'(x_k).$$

That concludes the proof of matching the conditions required.

The analytic procedure developed via the Lagrange polynomials is rather cumbersome. To overcome this, in [6] it is proposed to express the Hermite approximating polynomial in terms of the divided differences introduced in Section 1.1. In this form we write the Hermite approximating polynomial in the form of

$$p_H(x) = h_0 + h_1(x-x_0) + h_2(x-x_0)^2 + h_3(x-x_0)^2(x-x_1) + h_4(x-x_0)^2(x-x_1)^2$$
$$+ \ldots + h_{2n+1}(x-x_0)^2(x-x_1)^2 \cdots (x-x_{n-1}^2)(x-x_n).$$

The coefficients are discussed in the following computational example. This formulation is used in computer implementations, as the coefficients may be recursively computed. Finally, the error of the Hermite approximation is

$$f(x) - p_H(x) = \frac{(x-x_0)^2 \cdots (x-x_n)^2}{(2n+2)!} f^{(2n+2)}(\xi),$$

which has obvious similarities to the earlier error forms of both Newton and Lagrange.

1.3.1 Computational example

We consider the following simple example for demonstrating Hermite approximation. The three example points are actually obtained by sampling the $sin(x)$ curve from 0 to π. The points and the derivatives are shown in Table 1.2.

TABLE 1.2
Hermite approximation
example data

i	x_i	$f(x_i)$	$f'(x_i)$
0	0	0	1
1	$\pi/2$	1	0
2	π	0	-1

For three points the divided differences based formulation is

$$p_H(x) = h_0 + h_1(x-x_0) + h_2(x-x_0)^2 + h_3(x-x_0)^2(x-x_1)$$
$$+ h_4(x-x_0)^2(x-x_1)^2 + h_5(x-x_0)^2(x-x_1)^2(x-x_2).$$

The computation of the coefficients is facilitated by Table 1.3. Note the simplified notation for

$$f[z_{i,k}] = f[z_i, z_{i+1}, \ldots, z_{i+k}],$$

the alternating use of the divided difference and the first derivative in the $f[z_i, 1]$ column as well as the special setup of the starting columns.

TABLE 1.3
Hermite approximation computation

i	z_i	$f(z_i)$	$f[z_{i,1}]$	$f[z_{i,2}]$	$f[z_{i,3}]$	$f[z_{i,4}]$
0	0	0				
1	0	0	$f'(x_0) = 1$			
2	$\pi/2$	1	$2/\pi$	$4/\pi^2 - 2/\pi$		
3	$\pi/2$	1	$f'(x_1) = 0$	$-4/\pi^2$	$-16/\pi^3 + 4/\pi^2$	
4	π	0	$-2/\pi$	$-4/\pi^2$	0	$16/\pi^4 - 4/\pi^3$
5	π	0	$f'(x_2) = -1$	$4/\pi^2 - 2/\pi$	$16/\pi^2 - 4/\pi^2$	$16/\pi^4 - 4/\pi^3$

The top term from each column, starting from the third, gives the Hermite polynomial coefficients. With these, the Hermite approximation polynomial is

$$p_H(x) = f(x_0) + f'(x_0)(x - x_0) + f[z_{0,2}](x - x_0)^2 + f[z_{0,3}](x - x_0)^2(x - x_1)$$
$$+ f[z_{0,4}](x - x_0)^2(x - x_1)^2 + f[z_{0,5}](x - x_0)^2(x - x_1)^2(x - x_2).$$

We will utilize the fact that the term

$$f[z_{i,5}] = 0,$$

since the last two terms of the rightmost columns of Table 1.3 are identical. Numerically computing the coefficients and substituting yields

$$p_H(x) = 0 + 1(x - 0) - 0.23133(x - 0)^2 - 0.11076(x - 0)^2(x - \pi/2)$$
$$+ 0.035258(x - 0)^2(x - \pi/2)^2 + 0(x - 0)^2(x - \pi/2)^2(x - \pi).$$

We may collect the results in a quartic polynomial

$$p_H(x) = a_0 + a_1 x + a_2 x^2 + a_3 x^3 + a_4 x^4.$$

Figure 1.4 shows the approximation overlaid onto the $sin(x)$ function, from which the input data was sampled. One can observe an excellent matching of the points of the function inside the approximation interval, including the correct matching of the derivatives at the given points. Outside of the approximation interval the two curves separate, of course.

FIGURE 1.4 Hermite approximation example

1.4 Interpolation of functions of two variables with polynomials

Another natural extension of the interpolation problem is to approximate a function of two variables. This problem is much more generic and, depending on the distribution of the given dataset, it may not always be possible to solve it. We consider the problem given by

$$f(x_i, y_k), i = 0, 1, \ldots, m; k = 0, 1, \ldots, n.$$

This specifies a grid of $m + 1$ by $n + 1$ points in the $x - y$ plane as well as the corresponding function values. This is a very specific arrangement, as shown in Table 1.4, and it assures that the problem may be solved.

We are looking for a polynomial solution of the form

TABLE 1.4

Arrangement of
interpolation

$y \backslash x$	x_0	x_1	.	x_m
y_0	(f_{00})	f_{10}	.	$f_{m,0}$
y_1	(f_{01})	f_{11}	.	$f_{m,1}$
.	.	.	.	.
y_n	(f_{0n})	f_{1n}	.	$f_{m,n}$

$$p(x,y) = \sum_{i=0}^{m} \sum_{k=0}^{n} a_{ik} x^i y^k,$$

which satisfies

$$p(x_i, y_k) = f(x_i, y_k)$$

for all given points. This specifies a system of $(m+1)(n+1)$ equations and unknowns, and may be solved by any technique. It is, however, more efficient to solve this problem by a generalization of the Lagrange interpolation, which for this case may be formulated as

$$p(x,y) = \sum_{i=0}^{m} \sum_{k=0}^{n} f(x_i, y_k) L_i(x) L_k(y).$$

The proof of satisfying the interpolation conditions follows from the satisfaction of the x and y directional Lagrange base functions. The error of the approximation is

$$f(x,y) - p(x,y) = \frac{\omega_m(x)}{(m+1)!} \left[\frac{\partial^{m+1} f}{\partial x^{m+1}} \right]_{x=\xi_1}$$

$$+ \frac{\omega_n(y)}{(n+1)!} \left[\frac{\partial^{n+1} f}{\partial y^{n+1}} \right]_{y=\zeta_1} - \frac{\omega_m(x)}{(m+1)!} \frac{\omega_n(y)}{(n+1)!} \left[\frac{\partial^{m+1+n+1} f}{\partial x^{m+1} \partial y^{n+1}} \right]_{x=\xi_2, y=\zeta_2}.$$

Here the (ξ_1, ζ_1) and the (ξ_2, ζ_2) points are in the two-dimensional interval containing the interpolation points.

References

[1] Hermite, C.; Sur la formulae de Lagrange, *J. Reine Angew. Math.*, 1878.

[2] Kátai, I.; *Bevezetés a numerikus analizisbe*, Tankönyvkiadó, Budapest, 1975.

[3] Lagrange, J. L.; *Lecons Elémentaires sur les Mathématiques*, Ecole Normale, Paris, 1795.

[4] Newton, I.; *Philosophiae Naturalis Principia Mathematica*, London, 1687.

[5] Picard, E.; Sur la représentation approchée des fonctions, *C. R. Acad. Sci.*, Paris, 1891.

[6] Powell, M. J. D.; *Approximation Theory and Methods*, Cambridge University Press, Cambridge, 1981.

[7] *Tables of Lagrangian Interpolation Coefficients*, Columbia University Press, 1944.

[8] Wallis, J.; *Arithmetica Infinitorum*, Oxford, 1656.

[9] Weierstrass, K.; *Über the analytishe Darstellbarkeit sogenannter willkürlicher Funktionen einer reellen Veränderlichen*, Berliner Akad. Wiss., Berlin, 1885.

[10] Wendroff, B.; *Theoretical Numerical Analysis*, Academic Press, New York, 1966.

2

Approximation with splines

In the methods described in the last chapter, the order of the approximation polynomials was at least the same as the number of points. This is rather problematic when a large number of points is given, as the polynomials will have a large number of unnecessary undulations. In this chapter another class of approximations is introduced where the order of the approximation is kept to a small order, however large the number of points is. In order to overcome this conflict, a sequence of such polynomials is used and the technique is called spline approximation.

Spline is the name of the flexible rulers used in the not too distant past in the design of ships by naval architects; I also used them in the early 1970s as a junior engineer. They were made out of special tropical woods and were several meters in length. On the parquet floor of the draft rooms (sometimes called loft rooms) the low-scale drawings of the level curves of the ship were laid out with the help of the splines. The splines were held in place at certain points by supporting them with lead weights; otherwise the wood's natural flexibility defined the shape of the curves. Ship designers called the lead weights dolphins, a naming convention probably derived from the natural environment of their design objects.

This process has now been replaced by CAD systems. However, the mathematical tools used by such systems have their origin in the above mechanical tools. The mechanical model is used in deriving the formulae for the first type introduced here, the natural splines. Historically these are much more recent than the classical approximation tools, although one of the earliest references to natural splines [11] is about 60 years old.

The natural splines provided the foundation to surface approximations, as proposed by Coons [4] in the 1960s. The renaissance of the technology was dominated by the Bezier splines [3] and the B-splines [5], starting from the 1970s. These still constitute the foundation of most current computer-aided geometric modeling software.

2.1   Natural cubic splines

The natural cubic spline formulation presented here is inspired by the differential equation of the elastic bar, but other derivations are also known [1]. The bending of an elastic bar (the spline ruler) is described by

$$\frac{d^2y(x)}{dx^2} = \frac{M(x)}{IE},$$

where I is the cross-section moment of inertia and E is the Young's modulus of flexibility. As the bending moment is a continuous function of x, the solution curve $y(x)$ has second order continuity.

The problem of approximation is still posed in the form of a set of given points

$$(x_i, y_i); i = 0, 1, \ldots, n$$

with the assumption of $x_i < x_{i+1}$. The approximation function is going to be a piecewise function defined as

$$g(x) = \begin{cases} g_1(x), x_0 \leq x \leq x_1, \\ \qquad\qquad . \\ g_i(x), x_{i-1} \leq x \leq x_i, \\ \qquad\qquad . \\ g_n(x), x_{n-1} \leq x \leq x_n. \end{cases}$$

We now seek such segments of this piecewise approximation function that are thrice differentiable to satisfy the second order continuity between them. Such may be cubic segments of the form

$$g_i(x) = a_i x^3 + b_i x^2 + c_i x + d_i$$

for i values from 1 to n. The following three conditions must be satisfied for $i = 1, 2, \ldots, n-1$. First, the approximation functions must go through all the given points:

$$g_i(x_i) = g_{i+1}(x_i) = y_i.$$

Secondly, the tangents of the two polynomial segments from both sides of a point must be the same:

$$g_i'(x_i) = g_{i+1}'(x_i).$$

Finally, the second derivative (indicative of the curvature) of the functions at the connecting points must be equal:

$$g_i''(x_i) = g_{i+1}''(x_i) = r_i.$$

For the sake of simplicity we will assume that

$$r_0 = r_n = 0,$$

which implies zero curvature at the end points. This is not necessary, but it simplifies the discussion. This is called the natural end condition.

Since $g_i(x)$ is thrice differentiable, $g_i''(x)$ is at least linear. The linear approximation of the curvature between two points may be written in the form

$$g_i''(x) = r_{i-1}\frac{x_i - x}{h_i} + r_i\frac{x - x_{i-1}}{h_i},$$

where

$$h_i = x_i - x_{i-1} > 0.$$

Twice integrating and substituting the first and the third conditions yields

$$g_i(x) = r_{i-1}\frac{(x_i - x)^3}{6h_i} + r_i\frac{(x - x_{i-1})^3}{6h_i} + \frac{y_i - y_{i-1}}{h_i}x - \frac{r_i - r_{i-1}}{6}h_i x$$

$$+ \frac{x_i y_{i-1} - x_{i-1} y_i}{h_i} - \frac{x_i r_{i-1} - x_{i-1} r_i}{6}h_i$$

for $i - 1, 2, \ldots, n$. Note that the r_i curvatures are still not known. Differentiating this we get

$$g_i'(x) = \frac{-r_{i-1}(x_i - x)^2 + r_i(x - x_{i-1})^2}{2h_i} + \frac{y_i - y_{i-1}}{h_i} - \frac{r_i - r_{i-1}}{6}h_i.$$

Applying the second condition related to the first derivative, we finally obtain

$$h_i r_{i-1} + 2(h_i + h_{i+1})r_i + h_{i+1}r_{i+1} = 6\left(\frac{y_{i+1} - y_i}{h_{i+1}} - \frac{y_i - y_{i-1}}{h_i}\right)$$

for $i = 1, 2, \ldots, n - 1$. This is detailed as the following system of equations:

$$\begin{bmatrix} 2(h_1 + h_2) & h_2 & & \\ h_2 & 2(h_2 + h_3) & h_3 & \\ & & \ddots & \\ & & h_{n-1} & 2(h_{n-1} + h_n) \end{bmatrix} \begin{bmatrix} r_1 \\ r_2 \\ \vdots \\ r_{n-1} \end{bmatrix}$$

$$= 6 \begin{bmatrix} (y_2 - y_1)/h_2 - (y_1 - y_0)/h_1 \\ (y_3 - y_2)/h_3 - (y_2 - y_1)/h_2 \\ \vdots \\ (y_n - y_{n-1})/h_n - (y_{n-1} - y_{n-2})/h_{n-1} \end{bmatrix},$$

or in matrix form

$$HR = 6Y.$$

The H matrix is positive definite, so the system may always be solved. Substituting the now obtained r_i into the original $g_i(x)$ equation and executing a fair amount of algebra produces the coefficients of the natural spline.

$$a_i = \frac{r_i - r_{i-1}}{6h_i},$$

$$b_i = \frac{x_i r_{i-1} - x_{i-1} r_i}{2h_i},$$

$$c_i = \frac{r_i x_{i-1}^2 - r_{i-1} x_i^2}{2h_i} + \frac{y_i - y_{i-1}}{h_i} - \frac{r_i - r_{i-1}}{6} h_i,$$

and

$$d_i = \frac{x_i y_{i-1} - x_{i-1} y_i}{h_i} - \frac{x_i r_{i-1} - x_{i-1} r_i}{6} h_i + \frac{r_{i-1} x_i^3 - r_i x_{i-1}^3}{6h_i}.$$

2.1.1 Equidistant natural spline approximation

The equidistant case, always interesting to the engineer, also facilitates a significant simplification of the process. If $h_i = h$ for any $i = 1, \ldots, n$, then the equation simplifies to

$$\begin{bmatrix} 4 & 1 & & \\ 1 & 4 & 1 & \\ & \cdots & & \\ & & 1 & 4 \end{bmatrix} \begin{bmatrix} r_1 \\ r_2 \\ . \\ r_{n-1} \end{bmatrix} = \frac{6}{h^2} \begin{bmatrix} y_2 - 2y_1 + y_0 \\ y_3 - 2y_2 + y_1 \\ . \\ y_n - 2y_{n-1} + y_{n-2} \end{bmatrix}.$$

It is also possible to deviate from the natural end conditions and enforce either the tangents $g'(x_0), g'(x_n)$ or the curvatures $g''(x_0), g''(x_n)$ of the spline. The appropriate modifications are left to the reader.

On a final note, the desirable "smooth" behavior of the natural splines is of proven quality. Holladay's theorems states the following:

From the family of functions $G(x)$ that are at least twice differentiable in the interval $[x_0, x_n]$ and satisfy the $G''(x_0) = G''(x_n) = 0$ boundary conditions, the following functional,

$$\int_{x_0}^{x_n} [G''(x)]^2 dx,$$

is minimal if $G(x) = g(x)$, where $g(x)$ is the natural spline. The natural splines minimize the curvature of the approximating curve, a very desirable characteristic.

While the natural spline formulation was derived in terms of the approximation of planar curves, they are also possible to formulate in parametric form. That enables the technique to approximate space curves. The issue will be explored further in Section 2.4.1 in connection with the Coons patch.

2.1.2 Computational example

Let us again consider the points of

$$(x, y) = (4, 1); (6, 3); (8, 8); (10, 20),$$

but now to be approximated with a natural spline. Since the points are equidistant, the computational complexity will still be bearable by hand. Clearly

$$h_i = h = 2$$

and the system of equations to solve for the r_i is

$$\begin{bmatrix} 4 & 1 \\ 1 & 4 \end{bmatrix} \begin{bmatrix} r_1 \\ r_2 \end{bmatrix} = \frac{6}{4} \begin{bmatrix} 3 \\ 7 \end{bmatrix}.$$

The solution of $r_1 = 1/2$ and $r_2 = 5/2$ along with the boundary conditions $r_0 = 0$ and $r_3 = 0$ will enable us to find the polynomial coefficients for the three segments, shown in Table 2.1.

TABLE 2.1
Coefficients of a natural spline example

Segment	a	b	c	d
1	0.041667	−0.5	2.8333	−5.0
2	0.16667	−2.75	16.333	−32.0
3	−0.20833	6.25	−55.67	160.0

The segments of the natural spline curve, denoted by $s1, s2, s3$, are shown in Figure 2.1, demonstrate the curvature continuous connectivity between the neighboring spline segments. The first segment curve ($s1$) turns away after the second control point and, similarly, the second segment curve ($s2$) is below the curve prior to that point. Similar behavior can be observed with the third spline segment ($s3$); it is significantly above the spline between the first and the second points. The "points" are the input points in the figure.

This enables a comparison at the location $x = 5$ with the Lagrange interpolation. Here

$$g(5) = g_1(5) = a_1 5^3 + b_1 5^2 + c_1 5 + d_1 = 15/8,$$

which is identical to the value of $45/24$ obtained from both the Lagrange and Newton approximations.

Executing the example clearly demonstrates a computational shortcoming of the natural spline techniques: if one given point changes, the whole system

FIGURE 2.1 Natural spline interpolation polynomial

of equations needs to be resolved. In general the natural spline is a "global" spline; there is no way to locally change it without recomputing it.

2.2   Bezier splines

In order to obtain some local control over a spline, Bezier   [3] proposed a different approach. Given a set of points, the requirement of going through all points is released. The spline is required to go through the end points and some points in between. Some points, however, are only used to shape the curve.

Another difference from the last section is that we are going to develop the technique in parameter space. This also produces planar curves if needed, but generalizes to spatial curves easily. We are seeking the approximation polynomials in the parametric form

$$\underline{r}(t) = x(t)\underline{i} + y(t)\underline{j} + z(t)\underline{k}.$$

In the presentation here the points will still be approximated by a set of independent parametric cubic spline segments of the form

$$x(t) = a_x + b_x t + c_x t^2 + d_x t^3,$$
$$y(t) = a_y + b_y t + c_y t^2 + d_y t^3,$$

and

$$z(t) = a_z + b_z t + c_z t^2 + d_z t^3.$$

For simplicity of discussion, these will be commonly described as

$$S(t) = a + bt + ct^2 + dt^3,$$

where t ranges from 0.0 to 1.0.

First we focus on a single segment of the Bezier spline defined by four (control) points P_0, P_1, P_2, P_3, defining the Bezier polygon. We use the two intermediate points P_1, P_2, to define the starting and ending tangent lines of the curve. The distance of these middle points from the corner points will also influence the curve's shape. These four points define a Bezier [3] polygon, as shown in Figure 2.2.

As Figure 2.2 shows, the spline curve $f(x)$ is going through the corner points (P0= P_0, P3= P_3) but not the middle points (P1= P_1, P2= P_2). The Bezier spline segment is formed from four points as

$$S_B(t) = \sum_{i=0}^{3} P_i J_{3,i}(t).$$

Here

$$J_{3,i}(t) = \binom{3}{i} t^i (1-t)^{3-i}$$

are the Bernstein basis polynomials. Incidentally, Bernstein originally proposed such polynomials in his article [2] early in the last century while constructing a proof for Weierstrass' approximation theorem.

The role of the Bernstein polynomials here is "blending" the given point set into a continuous function, hence they are sometimes called the blending functions. Figure 2.3 shows the four Bernstein polynomials ($J_{3,i}$ denoted by Ji) and their shape helps in understanding the blending principle.

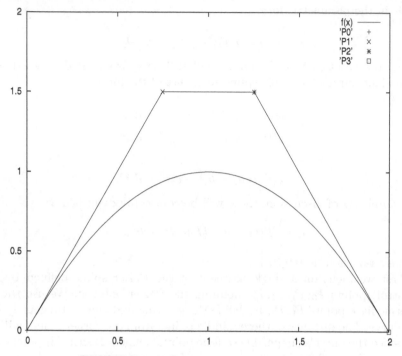

FIGURE 2.2 Bezier polygon and spline

Using the boundary conditions of the Bezier curve $(S_B(0), S_B(1), S'_B(0), S'_B(1))$ the matrix form of the Bezier spline segment may be written as

$$S_B(t) = TMP.$$

Here the matrix P contains the Bezier vertices

$$P = \begin{bmatrix} P_0 \\ P_1 \\ P_2 \\ P_3 \end{bmatrix},$$

and the matrix M the interpolation coefficients

$$M = \begin{bmatrix} 1 & 0 & 0 & 0 \\ -3 & 3 & 0 & 0 \\ 3 & -6 & 3 & 0 \\ -1 & 3 & -3 & 1 \end{bmatrix}.$$

FIGURE 2.3 Bernstein base polynomials

T is a parametric row vector:

$$T = \begin{bmatrix} 1 & t & t^2 & t^3 \end{bmatrix}.$$

Furthermore, introducing

$$S_B(t) = TA,$$

where

$$A = \begin{bmatrix} a \\ b \\ c \\ d \end{bmatrix},$$

it follows that

$$A = MP.$$

In detail,

$$\begin{bmatrix} a_x & a_y & a_z \\ b_x & b_y & b_z \\ c_x & c_y & c_z \\ d_x & d_y & d_z \end{bmatrix} = \begin{bmatrix} 1 & 0 & 0 & 0 \\ -3 & 3 & 0 & 0 \\ 3 & -6 & 3 & 0 \\ -1 & 3 & -3 & 1 \end{bmatrix} \begin{bmatrix} x_0 & y_0 & z_0 \\ x_1 & y_1 & z_1 \\ x_2 & y_2 & z_2 \\ x_3 & y_3 & z_3 \end{bmatrix}.$$

We will use this formula for the computation of the Bezier spline coefficients.

2.2.1 Rational Bezier splines

A very important generalization of this form is to introduce weight functions. The result is the rational parametric Bezier spline segment of the form

$$S_B(t) = \frac{\sum_{i=0}^{3} w_i P_i J_{3,i}(t)}{\sum_{i=0}^{3} w_i J_{3,i}(t)},$$

or in matrix notation,

$$S_B(t) = \frac{TM\overline{P}}{TMW}.$$

Here the vector

$$\overline{P} = \begin{bmatrix} w_0 P_0 \\ w_1 P_1 \\ w_2 P_2 \\ w_3 P_3 \end{bmatrix}$$

contains the weighted point coordinates and

$$W = \begin{bmatrix} w_0 \\ w_1 \\ w_2 \\ w_3 \end{bmatrix}$$

is the array of weights. The weights have the effect of moving the curve closer to the control points, P_1, P_2, shown as P1, P2 in Figure 2.4.

The location of a specified point on the curve P_s in Figure 2.4, define three values, in essence any three of the four weights. The remaining weight is defined by specifying the parameter value t^* to which the specified point should belong on the spline. Most commonly $t^* = \frac{1}{2}$ is chosen for such a point.

The three-dimensional approximation polynomial is described by

$$r_B(t) = \frac{TM\overline{X}}{TMW}\underline{i} + \frac{TM\overline{Y}}{TMW}\underline{j} + \frac{TM\overline{Z}}{TMW}\underline{k}.$$

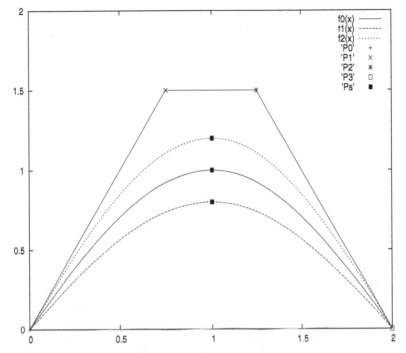

FIGURE 2.4 The effect of weights on the shape of a Bezier spline

Here

$$\overline{X} = \begin{bmatrix} w_0 x_0 \\ w_1 x_1 \\ w_2 x_2 \\ w_3 x_3 \end{bmatrix}, \; \overline{Y} = \begin{bmatrix} w_0 y_0 \\ w_1 y_1 \\ w_2 y_2 \\ w_3 y_3 \end{bmatrix}, \; \overline{Z} = \begin{bmatrix} w_0 z_0 \\ w_1 z_1 \\ w_2 z_2 \\ w_3 z_3 \end{bmatrix},$$

where x_i, y_i, z_i are the coordinates of the ith Bezier point.

An additional advantage of using rational Bezier splines [7] is to be able to exactly represent conic sections and quadratic surfaces. These are common components of industrial models, for manufacturing as well as aesthetic reasons.

In practical engineering problems there are likely to be many points and, therefore, a multitude of spline segments is required. The most important question arising in this regard is the continuity between segments, as shown in Figure 2.5.

FIGURE 2.5 Continuity of Bezier spline segments

Since the Bezier splines are always tangential to the first and last segments of the control polygon, clearly a first order continuity may exist only if the P_{i-1}, P_i, P_{i+1} points are collinear.

The presence of weights further specifies this continuity. Let us compute

$$\frac{\partial S_B}{\partial t}(t=0) = 3\frac{w_1}{w_0}(P_1 - P_0)$$

and

$$\frac{\partial S_B}{\partial t}(t=1) = 3\frac{w_2}{w_3}(P_3 - P_2),$$

where $(P_i - P_j)$ is a vector pointing from P_j to P_i. Focusing on the adjoining segments of spline $f(x)$ in Figure 2.5, the first order continuity condition is

$$\frac{w_{i-1}}{w_{i-0}}(P_i - P_{i-1}) = \frac{w_{i+1}}{w_{i+0}}(P_{i+1} - P_i).$$

There is a rather subtle but important distinction here. There is a geometric continuity component that means that the tangents of the neighboring spline segments are collinear. Then there is an algebraic component resulting in the fact that the magnitude of the tangent vectors is also the same. The notation w_{i+0}, w_{i-0} manifests the fact that the weights assigned to a control point in the neighboring segments do not have to be the same. If they are, a simplified first order continuity condition exists:

$$\frac{w_{i-1}}{w_{i+1}} = \frac{(P_{i+1} - P_i)}{(P_i - P_{i-1})}.$$

Enforcing such a continuity enhances the quality of the approximation. Furthermore, second order continuity is also possible. By definition,

$$\frac{\partial^2 S_B}{\partial t^2}(t = 0) = (6\frac{w_1}{w_0} + 6\frac{w_2}{w_0} - 18\frac{w_1^2}{w_0^2})(P_1 - P_0) + 6\frac{w_2}{w_0}(P_2 - P_1)$$

and

$$\frac{\partial^2 S_B}{\partial t^2}(t = 1) = (6\frac{w_1}{w_3} + 6\frac{w_2}{w_3} - 18\frac{w_2^2}{w_3^2})(P_2 - P_3) + 6\frac{w_1}{w_3}(P_1 - P_2).$$

Generalization to the boundary of neighboring segments, assuming that the weights assigned to the common point between the segments is the same, yields the second order continuity condition as

$$w_{i-2}(P_{i-2} - P_i) - 3\frac{w_{i-1}^2}{w_i}(P_{i-1} - P_i) = w_{i+2}(P_{i+2} - P_i) - 3\frac{w_{i+1}^2}{w_i}(P_{i+1} - P_i).$$

This is a rather strict condition requiring that the two control points prior and after the common point (five points in all) are coplanar with some additional constraints on the weight relations.

2.2.2   Computational example

We consider the point set $(0,0), (1,1), (2,1), (3,0)$. The matrix form introduced above produces

$$A = M \begin{bmatrix} 0 & 0 \\ 1 & 1 \\ 2 & 1 \\ 3 & 0 \end{bmatrix} = \begin{bmatrix} 0 & 0 \\ 3 & 3 \\ 0 & -3 \\ 0 & 0 \end{bmatrix}.$$

With these coefficients the Bezier spline for the given set of points is

$$S_B(t) = 3t\underline{i} + (3t - 3t^2)\underline{j}.$$

Let us now consider giving weights to the intermediate control points. The weight set of

$$W_2 = \begin{bmatrix} 1 \\ 2 \\ 2 \\ 1 \end{bmatrix}$$

will produce

$$\overline{P} = \begin{bmatrix} 0 & 0 \\ 2 & 2 \\ 4 & 2 \\ 3 & 0 \end{bmatrix}.$$

The resulting coefficients are

$$A_2 = M\overline{P} = \begin{bmatrix} 0 & 0 \\ 6 & 6 \\ 0 & -6 \\ -3 & 0 \end{bmatrix}.$$

The rational Bezier polynomial for this case is now written as

$$S_B(t) = \frac{6t - 3t^3}{1 + 3t - 3t^2}i + \frac{6t - 6t^2}{1 + 3t - 3t^2}j.$$

The results are organized into Table 2.2. The a, b, c, d coefficients are of the polynomial and the x,y subscripts indicate the spatial coordinates. The w subscript denotes the coefficients of the weight polynomial in the denominator.

TABLE 2.2
Coefficients of a Bezier spline example

weights	a_x, a_y	b_x, b_y	c_x, c_y	d_x, d_y	a_w	b_w	c_w	d_w
1	0,0	3,3	0,−3	0,0	1	0	0	0
2	0,0	6,6	0,−6	−3,0	1	3	−3	0
3	0,0	9,9	0,−9	−6,0	1	6	−6	0

The weights are the intermediate point weights, and the corner point weights are kept at unity. The first row contains the integral Bezier spline coefficients, the second row has intermediate weights of 2 and the third row increases the intermediate weights to 3. Figure 2.6 shows the resulting spline curves. Notice the tendency of the curve to move closer to the middle control points as the weights increase.

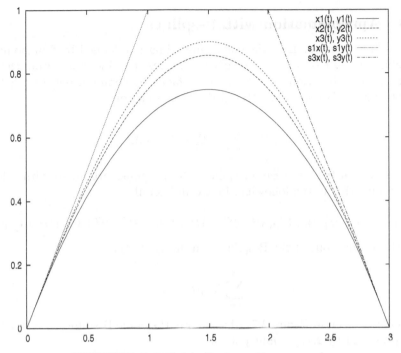

FIGURE 2.6 Cubic Bezier spline example

Throughout this chapter, the discussion was focused on producing piecewise cubic polynomials. The case of other than cubic approximations is briefly reviewed here. For example, using $m+1$ points, an mth order Bezier polynomial may be written as

$$S_B(t) = \sum_{i=0}^{m} P_i J_{m,i}(t).$$

Here

$$J_{m,i}(t) = \binom{m}{i} t^i (1-t)^{m-i}.$$

The most practical application of this is the case of $m = 2$, which, when used with weights, results in quadratic rational Bezier polynomials exactly reproducing conic sections. Higher than cubic polynomials are not often used with the Bezier approach.

2.3   Approximation with B-splines

Let us first focus again on the cubic approximation defined by four points. A B-spline segment is defined similarly to the Bezier spline, apart from the fact that the basis functions are different. In fact, the common belief is that the B-spline name is due to the "B" in the basis function

$$S_b(t) = \sum_{i=0}^{3} P_i B_{i,3}(t) = TM_bP.$$

To distinguish from the Bezier splines, the B-splines are noted with a lower case b subscript in the following. In specific detail,

$$S_b(t) = 1/6(1-t)^3 P_0 + 1/6(3t^3 - 6t^2 + 4)P_1 + 1/6(-3t^3 + 3t^2 + 3t + 1)P_2 + 1/6t^3 P_3.$$

The basis functions of the B-splines sum up to unity:

$$\sum_{i=0}^{3} B_{i,3}(t) = 1.$$

This can be seen in Figure 2.7, which shows the cubic B-spline basis functions ($B_{i,3}$ denoted by Bi) for four points.

Introducing matrix notation again,

$$S_b(t) = TM_bP,$$

where the M_b matrix of interpolating coefficients representing the blending functions is now

$$M_b = 1/6 \begin{bmatrix} 1 & 4 & 1 & 0 \\ -3 & 0 & 3 & 0 \\ 3 & -6 & 3 & 0 \\ -1 & 3 & -3 & 1 \end{bmatrix},$$

and the vector of the participating points

$$P = \begin{bmatrix} P_0 \\ P_1 \\ P_2 \\ P_3 \end{bmatrix},$$

as well as the vector T of parameters is the same as in the Bezier approach. The range of parameters is still from 0 to 1. This restriction will be released in the nonuniform formulation in the next section.

FIGURE 2.7 Cubic B-spline basis functions

2.3.1 Computational example

Using the same point set $(0,0), (1,1), (2,1), (3,0)$ as earlier produces

$$A_2 = M_b \begin{bmatrix} 0 & 0 \\ 1 & 1 \\ 2 & 1 \\ 3 & 0 \end{bmatrix} = \begin{bmatrix} 1 & 5/6 \\ 1 & 1/2 \\ 0 & -1/2 \\ 0 & 0 \end{bmatrix}.$$

Note that this is only the middle section of the B-spline approximation of the four-noded polygon. This section of the B-spline is described by

$$S_b, 1(t) = (t+1)\underline{i} + (-1/2t^2 + 1/2t + 5/6)\underline{j}.$$

This fact emphasizes the excellent local control of the B-splines. The four points contributed only to one section of the spline. Conversely, in order to change a section of a spline, only four points need to be modified.

The complete spline is plotted in Figure 2.8, where the ith section is marked by $xbi(t), ybi(t)$. Here "section" is used in contrast with segment. The latter is the complete spline defined by the four points and in turn multiple spline segments constitute a spline curve (when many points are given).

In order to find the first and third sections of the spline, another specific B-spline characteristic may be utilized, the duplication of control points. Duplicating the first control point gives rise to

$$A_1 = M_b \begin{bmatrix} 0 & 0 \\ 0 & 0 \\ 1 & 1 \\ 2 & 1 \end{bmatrix} = \begin{bmatrix} 1/6 & 1/6 \\ 1/2 & 1/2 \\ 1/2 & 1/2 \\ -1/6 & -1/3 \end{bmatrix}.$$

The duplication of the last control point results in

FIGURE 2.8 Cubic B-spline example

$$A_3 = M_b \begin{bmatrix} 1 & 1 \\ 2 & 1 \\ 3 & 0 \\ 3 & 0 \end{bmatrix} = \begin{bmatrix} 2 & 5/6 \\ 1 & -1/2 \\ 0 & -1/2 \\ -1/6 & 1/3 \end{bmatrix}.$$

Finally, forcing the spline through the beginning and the end points, another duplication of the end points is required:

$$A_0 = M_b \begin{bmatrix} 0 & 0 \\ 0 & 0 \\ 0 & 0 \\ 1 & 1 \end{bmatrix} = \begin{bmatrix} 0 & 0 \\ 0 & 0 \\ 0 & 0 \\ 1/6 & 1/6 \end{bmatrix},$$

and

$$A_4 = M_b \begin{bmatrix} 2 & 1 \\ 3 & 0 \\ 3 & 0 \\ 3 & 0 \end{bmatrix} = \begin{bmatrix} 17/6 & 1/6 \\ 1/2 & -1/2 \\ -1/2 & 1/2 \\ 1/6 & -1/6 \end{bmatrix}.$$

The complete spline segment spanning the four-point polygon consists of five sections. Note that in each of the sections, the parameter t ranged from 0 to 1, hence this is properly called a uniform B-spline.

2.3.2   Nonuniform B-splines

The nonuniform B-splines enable a different parameter distribution as well as basis function set for each curve section. This is based on a generic interpretation and computation of the basis functions.

$$B_{i,k}(t), i = 0, 1, \ldots, n; k = 0, 1, \ldots, m.$$

Note the distinction between the number of points used (n) and the order of the polynomial produced (m). In essence this enables the creation of various order spline segments from the same number of points.

The nonuniform B-spline basis functions are described recursively in terms of lower order functions. They are initialized as

$$B_{i,0}(t) = \begin{cases} 1, t_i \le t < t_{i+1}, \\ 0, t < t_i, t \ge t_{i+1}. \end{cases}$$

The higher order terms ($k = 1, 2, \ldots m$) are recursively formulated as

$$B_{i,k}(t) = \frac{t - t_i}{t_{i+k} - t_i} B_{i,k-1}(t) + \frac{t_{i+1+k} - t}{t_{i+1+k} - t_{i+1}} B_{i+1,k-1}(t).$$

For example, the basis functions for a cubic ($m = 3$) nonuniform B-spline are

$$B_{i,0}(t) = \begin{cases} 1, t_i \leq t < t_{i+1}, \\ 0, t < t_i, t \geq t_{i+1}. \end{cases}$$

Then

$$B_{i,1}(t) = \frac{t - t_i}{t_{i+1} - t_i} B_{i,0}(t) + \frac{t_{i+2} - t}{t_{i+2} - t_{i+1}} B_{i+1,0}(t),$$

$$B_{i,2}(t) = \frac{t - t_i}{t_{i+2} - t_i} B_{i,1}(t) + \frac{t_{i+3} - t}{t_{i+3} - t_{i+1}} B_{i+1,1}(t),$$

and

$$B_{i,3}(t) = \frac{t - t_i}{t_{i+3} - t_i} B_{i,2}(t) + \frac{t_{i+4} - t}{t_{i+4} - t_{i+1}} B_{i+1,2}(t).$$

This is repeated for $i = 0, 1 \ldots, n$. For example, given four points ($n = 3$), the parameter values range from $t_i = t_0$ to $t_{i+4} = t_7$. This accounts for eight parameter values, two for the intermediate points and two times three for the repeated segment boundaries.

In order to evaluate the recursive expressions of the basis functions when identical points are present, a special convention is needed. Specifically, when the denominator is zero, then the result of the expression is also considered to be zero.

These basis functions depend on the given nonuniform parametric intervals $[t_i, t_{i+1}]$. The t_i values are the parameter values defining the range of the parameters for the curve. They are presented in the knot vector. For example,

$$K = \begin{bmatrix} 0 & 1 & 2 & 3 \end{bmatrix}$$

defines the parameter range from 0 to 3 for the curve. The knot values of the end points need to be repeated, as shown in the example, to cover the whole span and in order to use the recursion formula. For the above parameter distribution, the cubic recursive formula requires the end point knot values to be in triplicate as

$$K = \begin{bmatrix} 0 & 0 & 0 & 1 & 2 & 3 & 3 & 3 \end{bmatrix}.$$

This is appropriately increased for even higher orders. This mechanism enables the multiple use of interior control points with duplicating values in the knot vector. A duplicate interior knot value will represent a multiple vertex with zero length in between. It is clearly a tool to locally control the curve, without changing the control point set. With an appropriate level of multiplication, the curve can be pulled as close to the point as desired.

2.3.3   Nonuniform rational B-splines

The B-spline technology can further be extended by adding weights. This process is the same as described in connection with the Bezier splines and not detailed further here.

The general form of a nonuniform rational B-spline, commonly called NURBS, is

$$S(t) = \frac{\sum_{i=0}^{n} w_i P_i B_{i,k}(t)}{\sum_{i=0}^{n} w_i B_{i,k}(t)}.$$

It is clearly an elaborate computation, best executed on computers and the implementation requires some programming skills. It is not easily demonstrated by a computational example. One practical implementation technique to deal with the rational nature and the weights is to use homogeneous coordinates,

$$Q(t) = [X(t), Y(t), Z(t), W(t)].$$

One can then calculate the NURBS in homogeneous coordinates,

$$\overline{S}(t) = \sum_{i=0}^{n} Q_i B_{i,k}(t).$$

After this, the NURBS needs to be mapped back to the original 3-space. For more details of computational algorithms, see [6] and [12].

These curves have become the industry workhorse for the past few decades. Some attractive reasons are their high flexibility, locally controllable nature and nonuniform parameterization. Most importantly, however, they are invariant under rotation, scaling, translation and perspective transformation. This last characteristic makes NURBS the darling of the computer graphics industry [10].

2.4   Surface spline approximation

All three types of splines may be generalized to surfaces if we approximate functions of two variables in parameter space. Surfaces may be generated from natural splines via Coons' method [4], but the Bezier and B-spline methods are the most widely used in the industry. Finally, tridiagonal surface patches are of practical importance, for example, in finite element discretization, the topic of Chapter 12.

2.4.1   Coons surfaces

Let us first generalize the natural spline into a parametric form. Instead of
the $y = f(x)$ form of the earlier section, we can write a natural spline in terms
of a parameter t as

$$x = x(t) = a_x t^3 + b_x t^2 + c_x t + d_x,$$

and similarly

$$y = y(t) = a_y t^3 + b_y t^2 + c_y t + d_y,$$

where $0 \le t \le 1$. With this approach a general natural spline in three dimensions may also be written

$$\underline{r}_N(t) = x(t)\underline{i} + y(t)\underline{j} + z(t)\underline{k},$$

where

$$z(t) = a_z t^3 + b_z t^2 + c_z t + d_z.$$

In the following we omit the N subscript, denoting the natural spline, for
clarity's sake. Coons' method creates a surface patch bounded by four natural
spline segments arranged in a parametric coordinate system defined by axes
u, v. The u parametric axis is represented by

$$\underline{r}(u, 0),$$

and the v parametric axis by

$$\underline{r}(0, v).$$

The other two sides of the patch are

$$\underline{r}(u, 1)$$

and

$$\underline{r}(1, v).$$

These are denoted by "r_u0, r_0v" and "r_u1, r_1v" in Figure 2.9, respectively.
Here $0 \le u \le 1$ and $0 \le v \le 1$. The two pairs define the sides of a rectangular
surface patch. Coons' recommendation in [4] is to interpolate the surface with
two blending functions. They are to blend the surface between these boundaries and are of the generic form

$$b_0(t) = b_{0,3} t^3 + b_{0,2} t^2 + b_{0,1} t + b_{0,0},$$

and

$$b_1(t) = b_{1,3} t^3 + b_{1,2} t^2 + b_{1,1} t + b_{1,0}.$$

FIGURE 2.9 Coons surface patch definition

Note that the coefficients have not yet been defined, but Coons required that the blending functions satisfy the following criteria:

$$b_0(0) = 1, \; b_0(1) = 0,$$

and conversely,

$$b_1(0) = 0, \; b_1(1) = 1.$$

It is also required that

$$b_0(t) + b_1(t) = 1,$$

for $0 \le t \le 1$. The interpolated surface is then of the form

$$\underline{r}(u, v) = \begin{bmatrix} b_0(u) \; b_1(u) \end{bmatrix} \begin{bmatrix} \underline{r}(0, v) \\ \underline{r}(1, v) \end{bmatrix} + \begin{bmatrix} b_0(v) \; b_1(v) \end{bmatrix} \begin{bmatrix} \underline{r}(u, 0) \\ \underline{r}(u, 1) \end{bmatrix}$$

$$- \begin{bmatrix} b_0(u) \; b_1(u) \end{bmatrix} \begin{bmatrix} \underline{r}(0, 0) \; \underline{r}(0, 1) \\ \underline{r}(1, 0) \; \underline{r}(1, 1) \end{bmatrix} \begin{bmatrix} b_0(v) \\ b_1(v) \end{bmatrix}.$$

This approach suffices when there is only one cubic spline surface patch. In practice, however, most of the time multiple segment spline curves describe the boundaries and therefore multiple spline patches or composite surfaces are needed. In order to afford appropriate continuity between the adjacent surface patches, derivatives at the boundaries also need to be defined.

In order to also interpolate the derivatives across the neighboring surface patch boundaries, two more blending functions are defined. They are

$$d_0(t) = d_{0,3}t^3 + d_{0,2}t^2 + d_{0,1}t + d_{0,0},$$

and

$$d_1(t) = d_{1,3}t^3 + d_{1,2}t^2 + b_{1,1}t + d_{1,0}.$$

Coons' criteria for the derivative blending functions are

$$d_0(0) = d_0(1) = d_1(0) = d_1(1) = 0,$$

but

$$\dot{d}_0(0) = \dot{d}_1(1) = 1$$

and

$$\dot{d}_0(1) = \dot{d}_1(0) = 0.$$

With these and the derivatives

$$r_u(u,v) = \frac{\partial r(u,v)}{\partial u},$$

$$r_v(u,v) = \frac{\partial r(u,v)}{\partial v},$$

and

$$r_{uv}(u,v) = \frac{\partial^2 r(u,v)}{\partial u \partial v},$$

the first order (tangent) continuous Coons patch is of the form

$$r(u,v) = \begin{bmatrix} b_0(u) & b_1(u) & d_0(u) & d_1(u) \end{bmatrix} \begin{bmatrix} r(0,v) \\ r(1,v) \\ r_u(0,v) \\ r_u(1,v) \end{bmatrix}$$

$$+ \begin{bmatrix} b_0(v) & b_1(v) & d_0(v) & d_1(v) \end{bmatrix} \begin{bmatrix} r(u,0) \\ r(u,1) \\ r_v(u,0) \\ r_v(u,1) \end{bmatrix}$$

$$- \begin{bmatrix} b_0(u) & b_1(u) & d_0(u) & d_1(u) \end{bmatrix} \begin{bmatrix} r(0,0) & r(0,1) & r_v(0,0) & r_v(0,1) \\ r(1,0) & r(1,1) & r_v(1,0) & r_v(1,1) \\ r_u(0,0) & r_u(0,1) & r_{u,v}(0,0) & r_{u,v}(0,1) \\ r_u(1,0) & r_u(1,1) & r_{u,v}(1,0) & r_{u,v}(1,1) \end{bmatrix} \begin{bmatrix} b_0(v) \\ b_1(v) \\ d_0(v) \\ d_1(v) \end{bmatrix}.$$

It is important to point out that the 4 by 4 matrix above contains four distinct types of quantities. The upper left terms are the corner points coordinates and they are of course given. The lower left and upper right four terms are the derivatives at the corner points with respect to either parametric coordinate and as such may also be computed from the given boundary curves.

On the other hand, the lower right-hand corner contains terms that are not computable from the boundary curves. They would be obtained from the surface function that is not available. These terms, called the corner twist terms, are hence approximated by the engineer. Often these terms are zeroed out, resulting in a flat corner area of the surface patch.

The remaining issue is to find an appropriate set of blending functions. A commonly used set is based on the Hermite interpolation introduced in the last chapter where the points and the derivatives were common between segments. For a parametric curve of

$$p(t) = a_0 + a_1 t + a_2 t^2 + a_3 t^3,$$

the end points of the curve are

$$p(0) = a_0$$

and

$$p(1) = a_0 + a_1 + a_2 + a_3.$$

The derivatives at the end points are

$$\dot{p}(0) = a_1$$

and

$$\dot{p}(1) = a_1 + 2a_2 + 3a_3.$$

Now we can reformulate this parametric curve in terms of the end points and the tangents at the end points as

$$p(t) = p(0)(1 - 3t^2 + 2t^3) + p(1)(3t^2 - 2t^3) + \dot{p}(0)(t - 2t^2 + t^3) + \dot{p}(1)(-t^2 + t^3).$$

This form defines the blending functions, often called Hermite basis functions, as

$$b_0 = 1 - 3t^2 + 2t^3,$$
$$b_1 = 3t^2 - 2t^3.$$

The blending functions for the derivatives are

$$d_0 = t - 2t^2 + t^3$$

and
$$d_1 = -t^2 + t^3.$$

It is easy to verify that these satisfy the required criteria stated above. One can write the spline in terms of these functions as

$$p(t) = b_0 p(0) + b_1 p(1) + d_0 \dot{p}(0) + d_1 \dot{p}(1).$$

To adhere to the formulation used in the Bezier approach, this is also commonly written in terms of a matrix as

$$p(t) = \begin{bmatrix} 1 & t & t^2 & t^3 \end{bmatrix} \begin{bmatrix} 1 & 0 & 0 & 0 \\ 0 & 0 & 1 & 0 \\ -3 & 3 & -2 & -1 \\ 2 & -2 & 1 & 1 \end{bmatrix} \begin{bmatrix} p(0) \\ p(1) \\ \dot{p}(0) \\ \dot{p}(1) \end{bmatrix} = T M_H P_H,$$

where the subscript H stands for Hermite and is distinguished from the conceptually similar, but content wise different, components of the Bezier form.

Finally, when a set of $n + 1$ parametric curves in the u direction and a set of $m + 1$ in the v direction are given, they define $m \cdot n$ surface patches. The composite Coons spline surface for the simpler, zero order continuous case is written as

$$\underline{r}(u, v) = \sum_{i=0}^{n} b_i(u) \underline{r}(i, v) + \sum_{j=0}^{m} (b_j(v) \underline{r}(u, j)$$

$$- \sum_{i=0}^{n} \sum_{j=0}^{m} b_i(u) b_j(v) \underline{r}(i, j).$$

2.4.2 Computational example

Consider the example surface patch shown in Figure 2.10. The corner points are shown in Table 2.3 along with the corresponding local parameter assignments.

Let us assume that the boundary spline curves are as follows. The u direction boundaries are

$$\underline{r}(u, 0) = u\underline{i} + 1 \cdot \underline{j} + 0 \cdot \underline{k},$$

and

$$\underline{r}(u, 1) = u\underline{i} + 0 \cdot \underline{j} + 1 \cdot \underline{k}.$$

The v direction boundaries are

$$\underline{r}(0, v) = 0 \cdot \underline{i} + (1 - v^2) \cdot \underline{j} + v\underline{k},$$

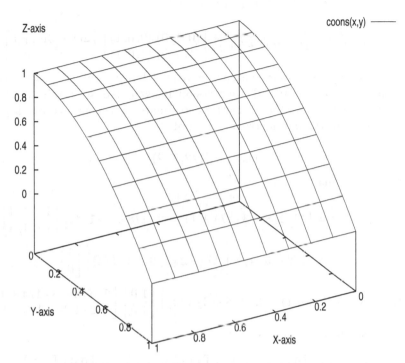

FIGURE 2.10 Coons surface patch example

TABLE 2.3
Coons surface patch
example

i	x_i	y_i	z_i	u	v
1	0	1	0	0	0
2	1	1	0	1	0
3	0	0	1	0	1
4	1	0	1	1	1

and
$$\underline{r}(1, v) = 1 \cdot \underline{i} + (1 - v^2) \cdot \underline{j} + v\underline{k}.$$

We will use the simpler, zero order continuous Coons patch and write the blending functions and the boundary curves into the formulation directly as

$$\underline{r}(u, v) = \left[(2u^3 - 3u^2 + 1) \; (-2u^3 + 3u^2) \right] \begin{bmatrix} 0 \cdot \underline{i} + (1 - v^2) \cdot \underline{j} + v\underline{k} \\ 1 \cdot \underline{i} + (1 - v^2) \cdot \underline{j} + v\underline{k} \end{bmatrix}$$

$$+ \left[(2v^3 - 3v^2 + 1)\, (-2v^3 + 3v^2) \right] \begin{bmatrix} u\underline{i} + 1 \cdot \underline{j} + 0 \cdot \underline{k} \\ u\underline{i} + 0 \cdot \underline{j} + 1 \cdot \underline{k} \end{bmatrix}$$

$$- \left[(2u^3 - 3u^2 + 1)\, (-2u^3 + 3u^2) \right] \begin{bmatrix} (0,1,0) & (0,0,1) \\ (1,1,0) & (1,0,1) \end{bmatrix} \begin{bmatrix} 2v^3 - 3v^2 + 1 \\ -2v^3 + 3v^2 \end{bmatrix}.$$

The analytic expansion of the patch is laborious, but possible. In practical surface spline approximations, the expression is evaluated at certain locations. For example, let us find the geometric coordinates of the surface patch point corresponding to the parametric location of

$$(u, v) = (1/2, 1/2).$$

For the y coordinate,

$$y(u, v) = 3/4 = \left[(2 \cdot 1/8 - 3 \cdot 1/4 + 1)\, (-2 \cdot 1/8 + 3 \cdot 1/4) \right] \begin{bmatrix} 1 - 1/4 \\ 1 - 1/4 \end{bmatrix}$$

$$+ \left[(2 \cdot 1/8 - 3 \cdot 1/4 + 1)\, (-2 \cdot 1/8 + 3 \cdot 1/4) \right] \begin{bmatrix} 1 \\ 0 \end{bmatrix}$$

$$- \left[(2 \cdot 1/8 - 3 \cdot 1/4 + 1)\, (-2 \cdot 1/8 + 3 \cdot 1/4) \right] \begin{bmatrix} 1 & 0 \\ 1 & 0 \end{bmatrix} \begin{bmatrix} 2 \cdot 1/8 - 3 \cdot 1/4 + 1 \\ -2 \cdot 1/8 + 3 \cdot 1/4 \end{bmatrix}.$$

Similarly for the x coordinate,

$$x(u, v) = \begin{bmatrix} 1/2 & 1/2 \end{bmatrix} \begin{bmatrix} 0 \\ 1 \end{bmatrix} + \begin{bmatrix} 1/2 & 1/2 \end{bmatrix} \begin{bmatrix} 1/2 \\ 1/2 \end{bmatrix} - \begin{bmatrix} 1/2 & 1/2 \end{bmatrix} \begin{bmatrix} 0 & 0 \\ 1 & 1 \end{bmatrix} \begin{bmatrix} 1/2 \\ 1/2 \end{bmatrix} = 1/2.$$

Finally, for the z coordinate,

$$z(u, v) = \begin{bmatrix} 1/2 & 1/2 \end{bmatrix} \begin{bmatrix} 1/2 \\ 1/2 \end{bmatrix} + \begin{bmatrix} 1/2 & 1/2 \end{bmatrix} \begin{bmatrix} 0 \\ 1 \end{bmatrix} - \begin{bmatrix} 1/2 & 1/2 \end{bmatrix} \begin{bmatrix} 0 & 1 \\ 0 & 1 \end{bmatrix} \begin{bmatrix} 1/2 \\ 1/2 \end{bmatrix} = 1/2.$$

The point sought on the surface is at

$$\underline{r}(u, v) = \underline{r}(1/2, 1/2) = \frac{1}{2}\underline{i} + \frac{3}{4}\underline{j} + \frac{1}{2}\underline{k}.$$

2.4.3 Bezier surfaces

In the surface approximation case, the approximating polynomial is defined by a set of points given in a rectangular array in parametric space. Such a case of control points and weights is shown in Figure 2.11. The rational parametric Bezier surface (patch) is described as

$$S_B(u, v) = \frac{\sum_{i=0}^{3} \sum_{j=0}^{3} w_{ij} J_{3,i}(u) J_{3,j}(v) P_{ij}}{\sum_{i=0}^{3} \sum_{j=0}^{3} w_{ij} J_{3,i}(u) J_{3,j}(v)}$$

FIGURE 2.11 Bezier surface patch weight definition

or in matrix form

$$S_B(u,v) = \frac{UM\overline{P}_{i,j}M^T V}{UMW_{i,j}M^T V}.$$

The computational components are

$$\overline{P}_{i,j} = \begin{bmatrix} w_{00}P_{00} & w_{01}P_{01} & w_{02}P_{02} & w_{03}P_{03} \\ w_{10}P_{10} & w_{11}P_{11} & w_{12}P_{12} & w_{13}P_{13} \\ w_{20}P_{20} & w_{21}P_{21} & w_{22}P_{22} & w_{23}P_{23} \\ w_{30}P_{30} & w_{31}P_{31} & w_{32}P_{32} & w_{33}P_{33} \end{bmatrix},$$

U is the parametric row vector of

$$U = \begin{bmatrix} 1 & u & u^2 & u^3 \end{bmatrix},$$

and for the other parameter

$$V = \begin{bmatrix} 1 \\ v \\ v^2 \\ v^3 \end{bmatrix}.$$

The weights now form a matrix of

$$W_{ij} = \begin{bmatrix} w_{00} & w_{01} & w_{02} & w_{03} \\ w_{10} & w_{11} & w_{12} & w_{13} \\ w_{20} & w_{21} & w_{22} & w_{23} \\ w_{30} & w_{31} & w_{32} & w_{33} \end{bmatrix}.$$

The approximating polynomial surface is described by

$$\underline{r}(u,v) = \frac{UM\overline{X}_{ij}M^T V}{UMW_{ij}M^T V}\underline{i} + \frac{UM\overline{Y}_{ij}M^T V}{UMW_{ij}M^T V}\underline{j} + \frac{UM\overline{Z}_{ij}M^T V}{UMW_{ij}M^T V}\underline{k}.$$

Here $\overline{X}_{ij}, \overline{Y}_{ij}, \overline{Z}_{ij}$, are the weighted point coordinates. Again, in practical circumstances a multitude of these patches is used to completely cover the data given. The earlier continuity discussion generalizes for surface patches. We consider the boundary between two patches, as shown in Figure 2.12. The partial derivative in the u direction at the boundary between two patches, represented by the slant arrow in the figure, may be written as

$$\frac{\partial S(u,v)}{\partial u}(i-0,j) = 3\frac{w_{i-1,j}}{w_{i,j}}(P_{i-1,j} - P_{i,j})$$

and

$$\frac{\partial S(u,v)}{\partial u}(i+0,j) = 3\frac{w_{i+1,j}}{w_{i,j}}(P_{i+1,j} - P_{i,j}).$$

The first order continuity condition is assured by

$$S_u^1 = \frac{\partial S(u,v)}{\partial u}(i-0,j) = \frac{\partial S(u,v)}{\partial u}(i+0,j).$$

From this it follows that

$$\frac{w_{i+1,j}}{w_{i-1,j}} = \frac{P_{i-1,j} - P_{i,j}}{P_{i+1,j} - P_{i,j}}.$$

Similarly

$$\frac{w_{i+1,j+1}}{w_{i-1,j+1}} = \frac{P_{i-1,j+1} - P_{i,j+1}}{P_{i+1,j+1} - P_{i,j+1}}.$$

As long as the above two conditions are satisfied, the first order continuity condition is satisfied along the u direction across the segment between the $P_{i,j}$

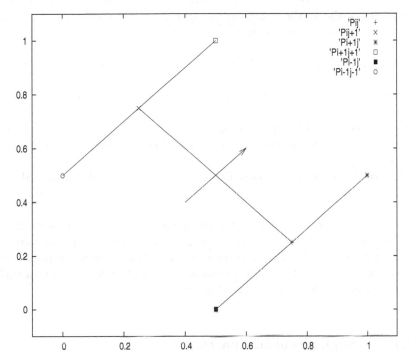

FIGURE 2.12 Bezier surface patch continuity

and $P_{i,j+1}$ points. A similar treatment is applied to the v parametric direction.

The second order continuity is based on the mixed second order partial derivatives of

$$\frac{\partial^2 S(u,v)}{\partial u \partial v}(i-0, j-0)$$

$$= 9\left(\frac{w_{i-1,j-1}}{w_{i,j}}(P_{i-1,j-1} - P_{i,j}) + \frac{w_{i-1,j-1}w_{i,j-1}}{w_{i,j}^2}(2P_{i,j} - P_{i-1,j} - P_{i,j-1})\right)$$

and

$$\frac{\partial^2 S(u,v)}{\partial u \partial v}(i+0, j+0)$$

$$= 9\left(\frac{w_{i+1,j+1}}{w_{i,j}}(P_{i+1,j+1} - P_{i,j}) + \frac{w_{i+1,j}w_{i,j+1}}{w_{i,j}^2}(2P_{i,j} - P_{i+1,j} - P_{i,j+1})\right).$$

Equivalencing these derivatives, as well as similar expressions with respect to the v parametric direction, results in the second order continuity condition.

This condition is almost overbearingly strict, requiring nine control points to be coplanar. Therefore, it is seldom enforced in engineering analysis. It mainly contributes to the aesthetic appearance of the surface created, and as such it is preferred by shape designers.

The Bezier objects are popular in the industry due to the following reasons:

1. All Bezier curves and surface patches are contained inside the convex hull of their control points.
2. The number of intersection points between a Bezier curve and an infinite plane is the same as the number of intersections between the plane and the control polygon.
3. All derivatives and products of Bezier functions are easily computed Bezier functions.

These properties are exploited in industrial geometric modeling computations. In industrial applications [9] spline curves are used to specify the faces of complex geometric models. This technique is called the boundary representation method, and it is widely used in various industries. A good overview of surface approximation methods is presented in [8].

2.4.4 Triangular surface patches

Both of the surface interpolation methods discussed above required the organization of input data in topologically rectangular arrangements. In real life geometries this may not always be possible, resulting in the need for a triangular surface element. This is especially important in finite element meshing technologies, where the most common mesh shape achieved by automated techniques is a triangle.

The triangular surface patch may be constructed with the help of areal or barycentric coordinates. Let us consider a triangle with vertices P_1, P_2 and P_3. The areal coordinates of a point P inside the triangle are written as

$$a_i = \frac{A_i}{A}, \ i = 1, 2, 3,$$

where A is the area of the triangle and A_i is the triangular area defined by P and $(P_j, j \neq i)$. For example, the a_1 areal coordinate is the area of the triangle defined by P, P_2, P_3, or in other words, opposite of P_1. Note that for all points on the edge opposite to corner P_i, the value of a_i is zero. It is clearly true that

$$\sum_{i=1}^{3} a_i = 1,$$

and any point inside the triangle may be written in terms of the areal coordinates as

$$P = \sum_{i=1}^{3} a_i P_i.$$

Hence the areal coordinates are truly the coordinates of P with respect to the corner points. We now introduce a vector of the areal coordinates

$$\underline{a} = (a_1, a_2, a_3),$$

and a vector of indices

$$\underline{i} = (i_1, i_2, i_3).$$

Here the i_1, i_2, i_3 indices are running on the edges of the triangle opposite to vertices P_1, P_2, P_3, respectively. With these, a method of generating triangular surface splines for the case

FIGURE 2.13 Triangular patch construction

TABLE 2.4

Triangular patch data

i_1	i_2	i_3	Point
3	0	0	$P_1 = Q_{300}$
0	3	0	$P_2 = Q_{030}$
0	0	3	$P_3 = Q_{003}$
2	1	0	Q_{210}
1	2	0	Q_{120}
2	0	1	Q_{201}
0	2	1	Q_{021}
1	0	2	Q_{102}
0	1	2	Q_{012}
1	1	1	Q_{111}

$$z = f(x, y) = f(\underline{a})$$

may be devised. We generalize the cubic Bernstein basis polynomials for this case as

$$B_{\underline{i}}(\underline{a}) = \frac{3!}{i_1! i_2! i_3!} a_1^{i_1} a_2^{i_2} a_3^{i_3}.$$

The evaluation of this expression follows the convention of $0! = 1$. Let the surface heights be given at the three vertices. Let the control point heights also be given at two intermediate points on each edge and at one point inside the triangle.

The control points of a triangular patch are shown in Figure 2.13 and their index arrangement is shown in Table 2.4. The given heights at the points are denoted by

$$z(\underline{i}) = f(Q_{i_1 i_2 i_3}).$$

Based on this the triangular surface is constructed as

$$z(\underline{a}) = \sum_{i_1 + i_2 + i_3 \leq 3} z(\underline{i}) B_{\underline{i}}(\underline{a}).$$

With the precomputed factorial expressions as coefficients

$$c_{\underline{i}} = \frac{3!}{i_1! i_2! i_3!},$$

the surface is interpolated as

$$z(\underline{a}) = \sum_{i_1 + i_2 + i_3 \leq 3} z(\underline{i}) c_{\underline{i}} a_1^{i_1} a_2^{i_2} a_3^{i_3}.$$

TABLE 2.5

Triangular patch computation

i_1	i_2	i_3	$c_{\underline{i}}$	$z(\underline{i})$	$a_1^{i_1}$	$a_2^{i_2}$	$a_3^{i_3}$
3	0	0	1	z_{300}	a_1^3	1	1
0	3	0	1	z_{030}	1	a_2^3	1
0	0	3	1	z_{003}	1	1	a_3^3
2	1	0	3	z_{210}	a_1^2	a_2	1
1	2	0	3	z_{120}	a_1	a_2^2	1
2	0	1	3	z_{201}	a_1^2	1	a_3
0	2	1	3	z_{021}	1	a_2^2	a_3
1	0	2	3	z_{102}	a_1	1	a_3^2
0	1	2	3	z_{012}	1	a_2	a_3^2
1	1	1	6	z_{111}	a_1	a_2	a_3

The computation process may be executed in the tabulated form of Table 2.5. The precomputed values greatly simplify the process, especially when there is a multitude of triangles.

2.4.5 Computational example

We generate the triangular patch shown in Figure 2.14, with the input data defined in Table 2.6. In the figure "side" marks the control polygon that partially coincides with the coordinate axes.

TABLE 2.6

Triangular patch example data

i_1	i_2	i_3	x	y	$z(\underline{i})$
3	0	0	0	0	0
0	3	0	1	0	1
0	0	3	0	1	1
2	1	0	1/3	0	0
1	2	0	2/3	0	1/3
2	0	1	0	1/3	0
1	0	2	0	2/3	1/3
0	1	2	1/3	2/3	1
0	2	1	2/3	1/3	1
1	1	1	1/3	1/3	0

FIGURE 2.14 Triangular patch example surface

TABLE 2.7
Triangular patch example computation

i_1	i_2	i_3	$c_{\underline{i}}$	$z(\underline{i})$	$a_1^{i_1}$	$a_2^{i_2}$	$a_3^{i_3}$
0	3	0	1	1	1	1/27	1
0	0	3	1	1	1	1	1/27
1	2	0	3	1/3	1/3	1/9	1
1	0	2	3	1/3	1/3	1	1/9
0	1	2	3	1	1	1/3	1/9
0	2	1	3	1	1	1/9	1/3

The computation for the surface point at $(x, y) = (1/3, 1/3)$ above the central control point is summarized in Table 2.7.

The areal coordinates for this point due to symmetry are

$$a_1 = a_2 = a_3 = \frac{1}{3},$$

since the area of the triangle is $1/2$. Note that only the rows producing nonzero results are shown in Table 2.7.

From the tabulated components the surface height at the desired location is

$$z(1/3, 1/3) = 2\frac{1}{27} + 3\frac{1}{3}(\frac{1}{3}\frac{1}{9} + \frac{1}{3}\frac{1}{9}) + 2 \cdot 3(\frac{1}{3}\frac{1}{9} + \frac{1}{9}\frac{1}{3}) = \frac{10}{27}.$$

The method demonstrated here generalizes to other than the $f(x, y)$ case shown above. There are also other ways to construct triangular patches. For example, one can use the rectangular representation and let two vertices coalesce, creating a degenerate rectangular patch. There are even more complex surface approximation shapes, like the n-sided surface patches described in [8].

References

[1] Ball, A. A.; A simple specification of the parametric cubic segment, *Journal of Computer Aided Design*, Vol. 10, pp. 181–182, 1978.

[2] Bernstein, S. N.; Sur la valeur asymtotique de la meilleure approximation des fonctions analytiques, *Comm. Russ. Acad. Sci. Paris*, Vol. 155, pp. 1062–1065, 1912.

[3] Bezier, P.; *Essai de definition numerique des courbes et de surfaces experimentals*, Universite D. et. M. Curie, Paris, 1977.

[4] Coons, S. A.; *Surfaces for Computer-Aided-Design of space forms*, MAC-TR-41, MIT, 1967.

[5] DeBoor, C.; On calculating with B-splines, *Journal of Approximation Theory*, Vol. 6, pp. 50–62, 1972.

[6] DeBoor, C.; *A Practical Guide to Splines*, Springer Verlag, New York, 1978.

[7] Farin, G. E.; Algorithms for rational Bezier curves, *Journal of Computer Aided Design*, Vol. 15, pp. 73–77, 1983.

[8] Gregory, J. A.; *The Mathematics of Surfaces*, Clarendon Press, Oxford, 1986.

[9] Komzsik, L.; *Computational Techniques of Finite Element Analysis*, CRC Press, Boca Raton, FL, 2005.

[10] Rogers, D. F. ed.; *State of the Art in Computer Graphics, Visualization and Modeling*, Springer Verlag, New York, 1991.

[11] Schoenberg, I. J.; Contributions to the problem of approximation of equidistant data by analytic functions, *Quarterly of Appl. Math.*, Vol. 4, pp. 45–99 and 112-141, 1946.

[12] Tiller, W.; Rational B-splines for curve and surface representation, *IEEE Journal of Computer Graphics*, Vol. 3, pp. 61–60, 1983.

3

Least squares approximation

This chapter concentrates on yet another approach to approximating a set of points. We first forced the approximation curve to go through all the points via the classical interpolation techniques. Then we somewhat relaxed this for the Bezier splines by only going through certain points with the spline approximations. Here, we are not going to go through the points at all, at least not intentionally.

Another noteworthy fact is that, while we had a measure of the error of the approximation in interpolation, we did not in the case of the splines. In the class of the least square methods, we go even further and try to minimize the error of the approximation.

The technique dates back to the 19th century. Legendre may have invented the concept in 1806 [4] and Gram has used the insight in [3] in 1883; however, earlier attempts were also made in similar contexts. The technique nowadays is widely described in most numerical analysis, approximation theory [2] and even calculus texts.

3.1 The least squares principle

We consider again a set of points

$$(x_i, y_i); i = 1, \ldots, n.$$

The least squares principle calls for an approximation function that produces the smallest sum of the square of the differences between the ordinates of the function and the given point set. Algebraically this error is

$$E_{LS} = \sum_{i=1}^{n} [y_i - LS(x_i)]^2,$$

where LS is the approximation function. The least squares principle requires this error to be minimal.

To obtain the minimum, the type of the approximating function must be chosen a priori. For a certain class of functions then, the equations defining the algebraic minimum will provide the computational strategy.

The most commonly used least squares approximation functions for a set of points are linear, polynomial and exponential functions. Their specific formulations will be detailed in the next sections. The more advanced and seldom published techniques of nonlinear, trigonometric and directional least squares methods will also be discussed. It is also notable that the principle may be applied in approximating functions. That is the topic of the next chapter.

3.2   Linear least squares approximation

The simplest and most widely used least squares approximation strategy is linear. In this case the approximation fits a straight line through the set of points, a line that minimizes the squared error.

Let us describe the linear approximation function as

$$LS(x) = ax + b.$$

The least squares error for such an approximating function is

$$E_{LS} = \sum_{i=1}^{n}(y_i - (ax_i + b))^2,$$

where the a, b coefficients of the approximating function are yet unknown. This error function is then a function of two variables and to minimize this, the first order partial derivatives must be zero.

$$\frac{\partial E_{LS}}{\partial a} = 0$$

and

$$\frac{\partial E_{LS}}{\partial b} = 0.$$

This would specifically give rise to

$$\frac{\partial}{\partial b}\sum_{i=1}^{n}[y_i - (ax_i + b)]^2 = -2\sum_{i=1}^{n}(y_i - ax_i - b) = 0$$

and

$$\frac{\partial}{\partial a} \sum_{i=1}^{n} [y_i - (ax_i + b)]^2 = -2 \sum_{i=1}^{n} (y_i - ax_i - b)(-x_i) = 0.$$

Evaluating the sums term wise results in a system of two equations in the two unknowns. These equations are called the normal equations:

$$bn + a \sum_{i=1}^{n} x_i = \sum_{i=1}^{n} y_i$$

and

$$b \sum_{i=1}^{n} x_i + a \sum_{i=1}^{n} x_i^2 = \sum_{i=1}^{n} x_i y_i.$$

The analytic solution of this system yields

$$b = \frac{\left(\sum_{i=1}^{n} x_i^2\right) \sum_{i=1}^{n} y_i - \left(\sum_{i=1}^{n} x_i y_i\right) \sum_{i=1}^{n} x_i}{n \sum_{i=1}^{n} x_i^2 - \left(\sum_{i=1}^{n} x_i\right)^2},$$

and

$$a = \frac{n \sum_{i=1}^{n} x_i y_i - \left(\sum_{i=1}^{n} x_i\right) \sum_{i=1}^{n} y_i}{n \sum_{i=1}^{n} x_i^2 - \left(\sum_{i=1}^{n} x_i\right)^2}.$$

The computation is rather straightforward, even with a large number of points. The application of such an approximation is very frequent when one tries to establish some trend in a large collection of data. Such collections occur, for example, when some engineering tests are executed and measured data is collected.

3.3   Polynomial least squares approximation

It is often very desirable to fit a polynomial to a set of points. In this case an a priori specification of the polynomial order is required. We consider an mth order polynomial of the form

$$LS_m(x) = a_m x^m + a_{m-1} x^{m-1} + \ldots + a_1 x + a_0.$$

Clearly, when $m = 1$ we have the case of the linear least squares approximation discussed in the last section. Following the footsteps of the linear process, the error is

$$E_{LS_m} = \sum_{i=1}^{n} (y_i - LS_m(x_i))^2.$$

Executing the square gives

$$E_{LS_m} = \sum_{i=1}^{n} y_i^2 - 2 \sum_{i=1}^{n} LS_m(x_i)y_i + \sum_{i=1}^{n} (LS_m(x_i))^2.$$

The extension of the sums is more complicated as we have $m+1$ terms. Some algebra yields

$$E_{LS_m} = \sum_{i=1}^{n} y_i^2 - 2 \sum_{j=0}^{m} a_j \sum_{i=1}^{n} x_i^j y_i + \sum_{j=0}^{m} \sum_{k=0}^{m} (a_j a_k \sum_{i=1}^{n} x_i^{j+k}).$$

In this case we have $m+1$ partial derivatives of the form

$$\frac{\partial E_{LS_m}}{\partial a_j} = -2 \sum_{i=1}^{n} x_i^j y_i + 2 \sum_{k=0}^{m} a_k \sum_{i=1}^{n} x_i^{j+k}.$$

Equating with zero and reordering produces the normal equations

$$\sum_{k=0}^{m} a_k \sum_{i=1}^{n} x_i^{j+k} = \sum_{i=1}^{n} x_i^j y_i; \quad j = 0, 1, \ldots, m.$$

The above represents a system of equations of $m+1$ unknowns.

$$a_0 \sum_{i=1}^{n} 1 + a_1 \sum_{i=1}^{n} x_i + \ldots + a_m \sum_{i=1}^{n} x_i^m = \sum_{i=1}^{n} y_i,$$

$$a_0 \sum_{i=1}^{n} x_i + a_1 \sum_{i=1}^{n} x_i^2 + \ldots + a_m \sum_{i=1}^{n} x_i^{m+1} = \sum_{i=1}^{n} x_i y_i,$$

and so on until the last equation of

$$a_0 \sum_{i=1}^{n} x_i^m + a_1 \sum_{i=1}^{n} x_i^{m+1} + \ldots + a_m \sum_{i=1}^{n} x_i^{2m} = \sum_{i=1}^{n} x_i^m y_i.$$

In matrix form this may be written as

$$\begin{bmatrix} n & \sum x_i & \cdots & \sum x_i^m \\ \sum x_i & \sum x_i^2 & \cdots & \sum x_i^{m+1} \\ \cdots & \cdots & \cdots & \cdots \\ \sum x_i^m & \sum x_i^{m+1} & \cdots & \sum x_i^{2m} \end{bmatrix} \begin{bmatrix} a_0 \\ a_1 \\ \cdots \\ a_m \end{bmatrix} = \begin{bmatrix} \sum y_i \\ \sum x_i y_i \\ \cdots \\ \sum x_i^m y_i \end{bmatrix},$$

or

$$A\underline{a} = \underline{b}.$$

Here the limits of summation were omitted to increase the readability. As the matrix is a Vandermonde matrix, this system of equations always has a unique solution when the x_i are all distinct.

The computations may be better organized by introducing the matrix

$$B = \begin{bmatrix} 1 & x_1 & \dots & x_1^m \\ 1 & x_2 & \dots & x_2^m \\ \dots & \dots & \dots & \dots \\ 1 & x_n & \dots & x_n^m \end{bmatrix}.$$

Then we compute

$$A = B^T B,$$

and

$$\underline{b} = B^T \underline{y}$$

where

$$\underline{y} = \begin{bmatrix} y_1 \\ y_2 \\ \dots \\ y_n \end{bmatrix}.$$

The solution is then

$$\underline{a} = A^{-1}\underline{b}.$$

3.4 Computational example

To demonstrate the computational process of the least squares technique, the point set of

$$(x, y) = (4, 1); (6, 3); (8, 8); (10, 20)$$

is used again. We will compute a linear and a quadratic least square approximation. The computation is facilitated by Table 3.1, from which the linear least squares approximation coefficients are found to be

$$a = 3.1,$$

and

$$b = -13.7.$$

This yields a linear approximation of

$$LS(x) = ax + b = 3.1x - 13.7.$$

TABLE 3.1

Computation of least squares approximation

x_i	y_i	x_iy_i	x_i^2	x_i^3	x_i^4	$x_i^2y_i$
4	1	4	16	64	256	16
6	3	18	36	216	1296	108
8	8	64	64	512	4096	512
10	20	200	100	1000	10000	2000
$\sum x_i$	$\sum y_i$	$\sum x_iy_i$	$\sum x_i^2$	$\sum x_i^3$	$\sum x_i^4$	$\sum x_i^2y_i$
28	32	286	216	1792	15648	2636

The approximation coefficients for the quadratic solution are based on the system

$$4a_0 + 28a_1 + 216a_2 = 32$$

$$28a_0 + 216a_1 + 1792a_2 = 286$$

$$216a_0 + 1792a_1 + 15648a_2 = 2636,$$

resulting in

$$a_2 = 0.625, a_1 = -5.65, a_0 = 13.8.$$

This yields

$$LS_2(x) = 0.625x^2 - 5.65x + 13.8.$$

The first and second order approximations are shown in Figure 3.1 along with the points. Note the significant difference in quality of the linear and the quadratic approximations. The computed least squares errors are

$$E_{LS} = \sum_{i=1}^{n}(y_i - LS(x_i))^2 = 25.8,$$

and

$$E_{LS_2} = \sum_{i=1}^{n}(y_i - LS_2(x_i))^2 = 0.8,$$

where the numerical superiority of the second order is very clear.

3.5 Exponential and logarithmic least squares approximations

Often a set of test data represents an analytic phenomenon of exponential or logarithmic behavior, as occurs, for example, in the engineering area of material sciences. In the first case it is desirable to compute an exponential least squares approximation as

$$LS_e(x) = ae^{bx}.$$

Here again, the a, b are the unknowns. The least squares principle dictates

$$E_{LS_e} = \sum_{i=1}^{n}(y_i - ae^{bx_i})^2$$

FIGURE 3.1 Least squares approximation example

to be minimum. The partial derivatives of this case are

$$\frac{\partial E_{LS_e}}{\partial a} = 2\sum_{i=1}^{n}(y_i - ae^{bx_i})(-e^{bx_i}),$$

and

$$\frac{\partial E_{LS_e}}{\partial b} = 2\sum_{i=1}^{n}(y_i - ae^{bx_i})(-ax_ie^{bx_i}).$$

Unfortunately, the resulting system of equations in unknowns a, b cannot be solved analytically. A well-known way to cope with this difficulty is to consider the logarithm of the approximation function:

$$ln(LS_e(x)) = ln(a) + bx.$$

This is clearly a linear least squares problem in $ln(a)$ and b and may be solved easily. It is important to note, however, that the solution obtained this way is not the least squares approximation of the exponential problem. It is a linear approximation of the $(x_i, ln(y_i))$ set of points. This solution may not be the best approximation of the (x_i, y_i) data set.

It is better to solve such a problem by solving the resulting system of nonlinear equations with some approximate methods. Such methods will be presented in Chapter 9.

Assuming now a logarithmic target function of

$$LS_l(x) = a + b \cdot ln(x),$$

the now familiar strategy yields the coefficients as

$$a = \frac{\sum_{i=1}^{n} y_i - b\sum_{i=1}^{n} ln(x_i)}{n}$$

and

$$b = \frac{n\sum_{i=1}^{n}(y_i ln(x_i)) - (\sum_{i=1}^{n} y_i)\sum_{i=1}^{n} ln(x_i)}{n\sum_{i=1}^{n} ln^2(x_i) - (\sum_{i=1}^{n} ln(x_i))^2}.$$

3.6   Nonlinear least squares approximation

The technique is very powerful when applied to nonlinear functions [1]. We assume a nonlinear function of the form

$$y = f(x, p_1, p_2, \ldots, p_m),$$

where $p_i, i = 1, 2, \ldots, m < n$ are parameters. We assume n data points given as

$$y_1 = f(x_1, p_1, p_2, \ldots, p_m),$$
$$y_2 = f(x_2, p_1, p_2, \ldots, p_m),$$

and so on, until

$$y_n = f(x_n, p_1, p_2, \ldots, p_m).$$

We seek the values of the yet unknown parameters of the function to minimize

$$z_i = y_i - f(x_i, p_1, p_2, \ldots, p_m)$$

for each $i = 1, 2, \ldots, n$. The rate of change of the distance is

$$dz_i = \sum_{j=1}^{m} \frac{\partial f}{\partial p_j}(x_i) dp_j$$

for each $i = 1, 2, \ldots, n$. Then in matrix form

$$d\underline{z} = A d\underline{p},$$

where

$$A = \begin{bmatrix} \frac{\partial f}{\partial p_1}(x_1) & \frac{\partial f}{\partial p_2}(x_1) & \cdots & \frac{\partial f}{\partial p_m}(x_1) \\ \frac{\partial f}{\partial p_1}(x_2) & \frac{\partial f}{\partial p_2}(x_2) & \cdots & \frac{\partial f}{\partial p_m}(x_2) \\ \cdots & \cdots & \cdots & \cdots \\ \frac{\partial f}{\partial p_1}(x_n) & \frac{\partial f}{\partial p_2}(x_n) & \cdots & \frac{\partial f}{\partial p_m}(x_n) \end{bmatrix},$$

$$d\underline{z} = \begin{bmatrix} dz_1 \\ dz_2 \\ \cdots \\ dz_n \end{bmatrix},$$

and

$$d\underline{p} = \begin{bmatrix} dp_1 \\ dp_2 \\ \cdots \\ dp_m \end{bmatrix}.$$

To produce a square system, we premultiply by A^T and obtain

$$A^T d\underline{z} = A^T A d\underline{p}.$$

Introduce an initial set of parameters

$$\underline{p}^0 = \begin{bmatrix} p_1^0 \\ p_2^0 \\ \cdots \\ p_n^0 \end{bmatrix},$$

and compute for each $i = 1, 2, \ldots, n$:

$$\Delta z_i^0 = y_i - f(x_i, p_1^0, p_2^0, \ldots, p_m^0).$$

These terms would be gathered into the array

$$\Delta \underline{z}^0 = \begin{bmatrix} \Delta z_1^0 \\ \Delta z_2^0 \\ \cdots \\ \Delta z_n^0 \end{bmatrix}.$$

Then, by using these differences instead of the differentials, the ensuing iteration process consisting of steps

$$\Delta \underline{p}^k = (A^T A)^{-1} A^T \Delta \underline{z}^{k-1},$$

and

$$p_j^k = p_j^{k-1} + \Delta \underline{p}^k(j)$$

will provide a better approximation for the parameters. Finally,

$$\Delta z_i^k = y_i - f(x_i, p_1^k, p_2^k, \ldots, p_n^k)$$

is the basis for the next iteration, if necessary, as well as a measure of convergence. The steps are executed until desirable accuracy measured by

$$||\Delta \underline{z}^k|| \leq \epsilon$$

is achieved, with a small ϵ.

The key to this technique is the easy computability of the partial derivatives of the function with respect to the parameters. A practical application for this technique is to approximate statistical data to some distribution function having variable parameters. For example, approximating measured data with a Gaussian distribution function of yet unknown mean value and standard deviation is an application for this technique.

3.6.1   Computational example

Another class of approximation functions well suited for this approach is the trigonometric class. We consider for this example the target function of

$$f(x) = p_1 sin(p_2 x),$$

with two unknown parameters, one for amplitude and another one for period. The input data of three points (just enough for the two parameters to make it a least squares solution) is

$$(x, y) = (0,0), (1/2, 5/4), (1,0).$$

The selection is admittedly very simple to enable easy computation while demonstrating the method. We will assume an initial parameter distribution of

$$p_1 = 1, p_2 = \pi,$$

or

$$\underline{p}^0 = \begin{bmatrix} 1 \\ \pi \end{bmatrix}.$$

The initial parameter assumption is quite reasonable as it defines the basic sine function. The derivatives are

$$\frac{\partial f}{\partial p_1} = sin(p_2 x),$$

and

$$\frac{\partial f}{\partial p_2} = p_1 \cdot x cos(p_2 x).$$

The A matrix is built as

$$A = \begin{bmatrix} sin(0 \cdot p_2) & 0 \cdot p_1 cos(0 \cdot p_2) \\ sin(p_2/2) & \frac{p_1}{2} cos(p_2/2) \\ sin(1 \cdot p_2) & p_1 cos(1 \cdot p_2) \end{bmatrix} - \begin{bmatrix} 0 & 0 \\ 1 & 0 \\ 0 & -1 \end{bmatrix}.$$

The starting vector for the iteration process is

$$\Delta \underline{z}^0 = \begin{bmatrix} 0 - 1 \cdot sin(0 \cdot \pi) \\ 5/4 - 1 \cdot sin(\pi/2) \\ 0 - 1 \cdot sin(\pi) \end{bmatrix} = \begin{bmatrix} 0 \\ 1/4 \\ 0 \end{bmatrix}.$$

Since for our specific example,

$$A^T A = \begin{bmatrix} 1 & 0 \\ 0 & 1 \end{bmatrix} = (A^T A)^{-1},$$

the first parameter adjustment is

$$\Delta \underline{p}^1 = \begin{bmatrix} 0 & 1 & 0 \\ 0 & 0 & -1 \end{bmatrix} \Delta \underline{z}^0 = \begin{bmatrix} 1/4 \\ 0 \end{bmatrix}.$$

With this, the adjusted parameter vector becomes

$$\underline{p}^1 = \underline{p}^0 + \Delta \underline{p}^1 = \begin{bmatrix} 5/4 \\ \pi \end{bmatrix}.$$

The corresponding solution adjustment is

$$\Delta \underline{z}^1 = \begin{bmatrix} 0 \\ 0 \\ 0 \end{bmatrix}.$$

This clearly indicates that we have reached the analytic solution of

$$f(x) = \frac{5}{4}sin(\pi x).$$

Varying the starting value for the second parameter will of course produce more iterations and less comfortable algebra, but work nevertheless.

3.7   Trigonometric least squares approximation

Periodic functions constitute a very important class in engineering work. The idea of least squares approximation of periodic data arises quite naturally. Let us consider a set of points with periodic tendencies and confined to the interval

$$-\pi = x_0 \leq x_1 \leq \ldots \leq x_i \leq \ldots \leq x_{2n} = \pi.$$

This specific interval restriction may always be overcome by a simple coordinate transformation. Assuming the point set is given in the interval from 0 to t, the

$$\overline{x}_i = \pi \frac{x_i - \frac{t}{2}}{\frac{t}{2}}$$

transformation will suffice. The matching ordinate values then are

$$f(x_i) = f(\frac{t}{2\pi}\overline{x}_i + \frac{t}{2}).$$

An equidistant point distribution, while making the computations simpler, is not a requirement for the technique developed below.

We are seeking a periodic or trigonometric least squares approximation of the data in the form of

$$LS_p(x) = \frac{a_0}{2} + \sum_{k=1}^{m}(a_k cos(kx) + b_k sin(kx)).$$

We will set $b_m = 0$ and use the form

$$LS_p(x) = \frac{a_0}{2} + a_m cos(mx) + \sum_{k=1}^{m-1}(a_k cos(kx) + b_k sin(kx)).$$

Since we need to find $2m$ coefficients, the number of points $2n + 1$ must be greater; otherwise the problem cannot be solved.

This formulation, which will be discussed at length in the next chapter in connection with approximating functions, is also called the discrete Fourier approximation. Here we apply it to discrete data in a least squares approximation sense. Because the 2nth point is located at $x_{2n} = \pi$, for which $sin(x_{2n}) = 0$, we are considering only $2n - 1$ points in defining the least squares error and that is why we set b_m to zero.

$$E_{LS_p} = \sum_{i=0}^{2n-1} (y_i - LS_p(x_i))^2.$$

The equations defining the approximation coefficients are derived again from the stationary values of the first partial derivatives of the error function:

$$\frac{\partial E_{LS_p}}{\partial a_k} = 0$$

and

$$\frac{\partial E_{LS_p}}{\partial b_k} = 0.$$

For the coefficients in the sum, $k = 1, 2 \ldots m - 1$, these equations read

$$2 \sum_{i=0}^{2n-1} (y_i - LS_p(x_i))(-cos(kx_i)) = 0,$$

and

$$2 \sum_{i=0}^{2n-1} (y_i - LS_p(x_i))(-sin(kx_i)) = 0.$$

The derivatives with respect to the first and the last a coefficients are similar, but simpler.

$$\sum_{i=0}^{2n-1} (y_i - LS_p(x_i)) = 0,$$

and

$$2 \sum_{i=0}^{2n-1} (y_i - LS_p(x_i))(-cos(mx_i)) = 0.$$

After substituting, reordering and a considerable amount of algebraic work, the resulting approximation formulae are

$$b_k = \frac{1}{2n} \sum_{i=0}^{2n-1} y_i sin(kx_i); \quad k = 1, 2, \ldots m - 1,$$

and

$$a_k = \frac{1}{2n} \sum_{i=0}^{2n-1} y_i cos(kx_i); k = 0, 1, 2, \ldots m.$$

Note that the latter formula now includes the two specific a coefficients.

3.7.1   Computational example

Let us consider the set of points given in the first three columns of Table 3.2, clearly demonstrating periodic tendencies.

TABLE 3.2
Periodic least squares
example

i	x_i	y_i	$LS_p(x_i)$
0	$-\pi$	1	1
1	$-\pi/2$	-1	-1
2	0	1	1
3	$\pi/2$	-1	-1
4	π	1	1

Let us seek a trigonometric approximation of order $m = 2$ for this set of $2n+1 = 5$ points. Using the above formulae for the first a coefficient produces

$$a_0 = \frac{1}{4} \sum_{i=0}^{3} y_i cos(0x_i) = \frac{1}{4} \sum_{i=0}^{3} y_i = \frac{1}{4}(1 - 1 + 1 - 1) = 0.$$

The second a coefficient computes as

$$a_1 = \frac{1}{4} \sum_{i=0}^{3} y_i cos(x_i)$$

$$= \frac{1}{4}(cos(x_0) - cos(x_1) + cos(x_2) - cos(x_3)) = 0.$$

Similarly for the third $m = 2$:

$$a_2 = \frac{1}{4} \sum_{i=0}^{3} y_i cos(2x_i)$$

$$= \frac{1}{4}(cos(2x_0) - cos(2x_1) + cos(2x_2) - cos(2x_3)) = 1.$$

Finally the single b coefficient in this case also becomes zero as

$$b_1 = \frac{1}{4} \sum_{i=0}^{3} y_i sin(x_i)$$

$$= \frac{1}{4}(sin(x_0) - sin(x_1) + sin(x_2) - sin(x_3)) = 0.$$

The least squares approximation of the given dataset is obtained in the form as

$$LS_p(x) = cos(2x).$$

The approximation values at the input locations, shown in the last column of Table 3.2, exactly match the input data. This is not a surprising result as the input data was produced by sampling the $cos(2x)$ function at the specific discrete points.

3.8 Directional least squares approximation

The least squares concept introduced in and discussed throughout this chapter is based on the vertical distances between the given points and the approximating curve. It is of engineering interest on occasion to define the distance measurement of the approximation in a certain direction. An often used case of directional least squares methods is the case of perpendicular directions [5].

In this case, the distance between the line and the given point is measured perpendicularly to the line. This is of course the smallest distance from the point to the line. The distance is specifically

$$d_i = \frac{|y_i - (a + bx_i)|}{\sqrt{1 + b^2}}.$$

The least squares error function becomes

$$E_{LSp} = \sum_{i=1}^{n} \frac{(y_i - (a + bx_i))^2}{1 + b^2}.$$

The now very familiar conditions for the minimum are

$$\frac{\partial E_{LSp}}{\partial a} = \frac{2}{1 + b^2} \sum_{i=1}^{n} ((a + bx_i) - y_i) = 0$$

and

$$\frac{\partial E_{LSp}}{\partial b} = \frac{2}{1+b^2} \sum_{i=1}^{n}((a+bx_i)-y_i)x_i - \frac{2b}{(1+b^2)^2} \sum_{i=1}^{n}((a+bx_i)-y_i)^2 = 0.$$

The latter equation reduces to

$$(1+b^2)\sum_{i=1}^{n}((a+bx_i)-y_i)x_i - b\sum_{i=1}^{n}((a+bx_i)-y_i)^2 = 0,$$

which is unfortunately very convoluted. With some very laborious algebraic work [5] it actually reduces to a quadratic with respect to b.

$$b^2 + \frac{\sum y_i^2 - \sum x_i^2 + (\sum x_i)^2/n - (\sum y_i)^2/n}{(\sum x_i \sum y_i)/n - \sum x_i y_i} b - 1 = 0.$$

Here the limits of summation were again omitted to increase the readability. The former equation is linear with respect to a, so it is somewhat simpler to compute. The process is conceptually very easy, but algebraically cumbersome to extend to other than the perpendicular direction.

3.9   Weighted least squares approximation

Finally we briefly review another extension of the least squares principle, in which the input data is weighted. It is quite conceivable that the data gathering process has some level of uncertainty associated with it. This uncertainty may be related to the actual measurements; for example, it is possible that the larger the measured value, the bigger the measurement error is. This could be the case with some engineering measuring devices.

The issue may be addressed by assigning weights to the input data, accounting for the uncertainty of the measuring or gathering process. The resulting weighted linear least squares principle is to minimize

$$E_{wLS} = \sum_{i=1}^{n} w_i[y_i - LS(x_i)]^2.$$

The development of the solution following Section 3.1 is straightforward and not detailed here. The values of the weights are sometimes chosen on a statistical basis; the standard variation σ_i, for example, is often used as a weight.

References

[1] Bates, D. M. and Watts, D. G.; *Nonlinear Regression and Its Applications*, Wiley, New York, 1988.

[2] Cheney, E. W.; *Introduction to Approximation Theory*, McGraw-Hill, New York, 1966.

[3] Gram, J. P.; Über the Entwicklung reeller Funktionen in Reihen mittels der Methode der kleinsten Quadrate, *J. Reine Angew. Math.*, Vol. 94, pp. 41–73, 1883.

[4] Legendre, A, M.; *Nouvelles methodes pour la determination des orbites des cometes*, Courcier, Paris, 1806.

[5] Sardelis, D. and Valahas, T.; *Least squares fitting with perpendicular offsets*, Internet posting, 2003.

4

Approximation of functions

The underlying problem description of the prior three chapters was a set of points, even if they were obtained by sampling a function. In the following three chapters, the problem foundation will always be a function.

Legendre already had made his mark in this area by the end of the 18th century [9], and a class of polynomials bearing his name will be discussed in this chapter. The subject was also studied extensively by Chebyshev [3] in the middle of the 19th century. Trigonometric approximations attributed to Fourier are also quite old [4]. Rational approximation was pioneered by Padé at the end of the 19th century [10].

4.1   Least squares approximation of functions

To measure the approximation quality, the least squares principle introduced in the last chapter for a discrete set of points may be applied here also. We consider now a continuous function $f(x)$ given in the interval $[a \leq x \leq b]$. We are looking for an approximation function $g(x)$ that minimizes the following integral:

$$\int_a^b (f(x) - g(x))^2 dx = min.$$

The expression may be generalized to include a weight function $\rho(x)$ such that

$$\int_a^b \rho(x)(f(x) - g(x))^2 dx = min.$$

Such an approximating function may be conveniently constructed by a linear combination of functions

$$g(x) = \sum_{k=0}^{n} c_k g_k(x).$$

The linear dependency of the g_k component functions, sometimes called basis functions, is manifested in their Gramian matrix of scalar products:

$$G = \begin{bmatrix} (g_0, g_0) & (g_0, g_1) & \cdot & (g_0, g_n) \\ (g_1, g_0) & (g_1, g_1) & \cdot & (g_1, g_n) \\ \cdot & \cdot & \cdot & \cdot \\ (g_n, g_0) & (g_n, g_1) & \cdot & (g_n, g_n) \end{bmatrix}.$$

Here the scalar products are defined as

$$(g_i, g_j) = \int_a^b g_i(x)g_j(x)dx.$$

The system of basis functions is linearly independent if

$$det(G) \neq 0.$$

Without loss of generality, let us restrict the interval of approximation to $[-1 \leq x \leq 1]$. This may always be achieved easily by a linear transformation of

$$t = \frac{a+b}{2} + \frac{b-a}{2}.$$

Let us also assume, for now, unit weights

$$\rho(x) = 1.$$

Furthermore, let us require that the linearly independent system of functions is orthogonal. For this it is required that

$$(g_i, g_j) = \int_a^b g_i(x)g_j(x)dx = \begin{cases} 0, i \neq j, \\ ||g_k||^2, i = j. \end{cases}$$

Then the Gramian matrix is diagonal:

$$G = \begin{bmatrix} ||g_0||^2 & 0 & 0 & 0 \\ 0 & ||g_1||^2 & 0 & 0 \\ \cdot & \cdot & \cdot & \cdot \\ 0 & 0 & 0 & ||g_n||^2 \end{bmatrix}.$$

The approximation function of

$$g(x) = \sum_{k=0}^{n} c_k g_k(x)$$

with such a set of basis functions results in the system of equations

$$||g_k||^2 c_k = (f, g_k); k = 0, 1, \ldots, n,$$

which may be solved for the unknown coefficients as

$$c_k = \frac{(f, g_k)}{||g_k||^2}; k = 0, 1, \ldots, n.$$

This is even simpler if the basis functions are not just orthogonal, but orthonormal:

$$(g_i, g_j) = \int_a^b g_i(x) g_j(x) dx = \begin{cases} 0, i \neq j, \\ 1, i = j. \end{cases}$$

Then the coefficients are simply

$$c_k = (f, g_k); k = 0, 1, \ldots, n.$$

Such a set of basis functions may be constructed from the class of power functions. Precisely, the selection of

$$g_k(x) = x^k; k = 0, 1, \ldots, n$$

with different weight functions gives rise to various classes of orthogonal polynomial approximations that are detailed in the following sections.

4.2 Approximation with Legendre polynomials

We are focusing on the interval $[-1 \leq x \leq 1]$ and the use of the power basis functions with

$$\rho(x) = 1$$

results in the Legendre polynomial approximation. The members of the Legendre polynomials may be obtained in several ways. One way is to execute a Gram–Schmidt orthogonalization procedure [5] on the given class of basis polynomials above. The procedure will be detailed next.

4.2.1 Gram–Schmidt orthogonalization

The procedure is well known by engineers in a linear algebraic sense. However, here we use it in connection with a more generic scalar product. The algorithm of the procedure for our case is as follows. For

$$g_k(x) = x^k; k = 0, 1, \ldots, n,$$

compute

$$G_k = g_k - \sum_{j=0}^{k-1} \frac{(g_k, G_j)}{(G_j, G_j)} G_j.$$

Here

$$(g_k, G_j) = \int_{-1}^{+1} g_k(x) G_j(x) dx$$

and

$$(G_j, G_j) = \|G_j(x)\|^2.$$

The execution of this algorithm starts as

$$G_0 = g_0 = 1,$$

and

$$\|G_0\|^2 = \int_{-1}^{+1} 1 \cdot 1 dx = 2.$$

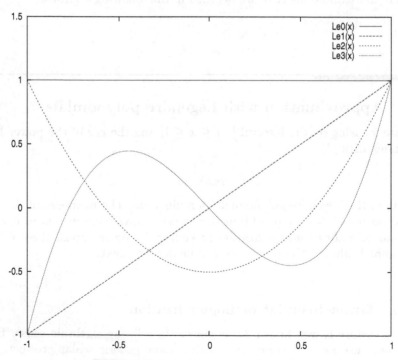

FIGURE 4.1 Legendre polynomials

The next term is computed as

$$G_1 = x - \frac{(x,1)}{2}1 = x - 0 = x,$$

$$\|G_1\|^2 = \int_{-1}^{+1} x \cdot x dx = 2/3.$$

Finally the third term is

$$G_2 = x^2 - \frac{(x^2,1)}{2}1 - \frac{(x^2,x)}{2/3}x = x^2 - 1/3 \cdot 1 - 0 \cdot x = x^2 - \frac{1}{3}.$$

The Legendre polynomials (noted by $Le_k(x)$) are obtained from these orthogonal polynomials via a specific normalization as

$$Le_k(1) = \alpha_k G_k(1) = 1.$$

Clearly,

$$Le_0(x) = G_0(x), Le_1(x) = G_1(x).$$

On the other hand,

$$\alpha_2 G_2(1) = 1$$

yields

$$\alpha_2 = \frac{3}{2},$$

and

$$Le_2(x) = \frac{3}{2}G_2(x) = \frac{3}{2}(x^2 - \frac{1}{3}) = \frac{1}{2}(3x^2 - 1).$$

The three members obtained so far enable us to construct a recurrence formula for higher order members. One can see that

$$2Le_2(x) - 3xLe_1(x) + 1Le_0(x) = 0.$$

It may be proven by induction that

$$kLe_k(x) - (2k - 1)xLe_{k-1}(x) + (k - 1)Le_{k-2}(x) = 0.$$

The next term from the equation is

$$Le_3(x) = \frac{1}{2}(5x^3 - 3x),$$

and so on. These first four Legendre polynomials are shown in Figure 4.1. It is visibly apparent and possible to prove that

$$Le_k(x) = (-1)^k Le_k(-x),$$

i.e., they are symmetric with respect to the y axis when k is even and about the origin when odd. The final goal of this section is the approximation of

$$f(x) = \sum_{k=0}^{n} c_k Le_k(x),$$

for which the coefficients are still outstanding. Since the Legendre polynomials are only orthogonal, not orthonormal, the scalar products of

$$(Le_k(x), Le_k(x)) = ||Le_k(x)||^2$$

are needed. These may be computed easily as

$$||Le_0||^2 = 2; \ ||Le_1||^2 = \frac{2}{3}; \ ||Le_2||^2 = \frac{2}{5}.$$

A pattern may again be recognized:

$$||Le_k||^2 = \frac{2}{2k+1},$$

which may be verified by

$$||Le_3||^2 = \frac{2}{7} = \frac{2}{2 \cdot 3 + 1}.$$

Hence, the approximation coefficients are

$$c_k = \frac{2k+1}{2}(f(x), Le_k(x)); \ k = 0, 1, \dots, n.$$

4.2.2 Computational example

To provide a working knowledge of the Legendre approximation process, we conclude this section with a simple computational example of a Legendre polynomial approximation of $f(x) = x^3$. There is, of course, no practical justification to approximate this function with Legendre polynomials, apart from the exercise's educational purpose.

Let us first compute the scalar products needed for the approximation coefficients:

$$(f(x), Le_0(x)) = (x^3, 1) = 0,$$

$$(f(x), Le_1(x)) = (x^3, x) = \frac{2}{5},$$

$$(f(x), Le_2(x)) = (x^3, \frac{1}{2}(3x^2 - 1)) = 0,$$

and

$$(f(x), Le_3(x)) = (x^3, \frac{1}{2}(5x^3 - 3x)) = \frac{4}{35}.$$

The two meaningful coefficients are

$$c_1 = 3/2 \cdot 2/5 = \frac{3}{5},$$

and

$$c_3 = 7/2 \cdot 4/35 = \frac{2}{5}.$$

The approximation via Legendre polynomials is

$$f(x) = c_1 Le_1(x) + c_3 Le_3(x) = \frac{3}{5}x + \frac{2}{5}\frac{1}{2}(5x^3 - 3x) = x^3.$$

This is, of course, the ultimate least squares fit for the given function, as the approximation function analytically agrees with the function to be approximated.

The Legendre polynomials have several other very interesting characteristics, for example; they may be derived as a solution of a specific differential equation. They, specifically their zeroes, will also have a prominent role in Chapter 6 in the Gaussian numerical quadrature process.

4.3   Chebyshev approximation

Similarly to the Legendre approximation, the Chebyshev approximation is based on the linearly independent set of

$$g_k(x) = x^k$$

power functions. Chebyshev, however, defines the scalar product in terms of the weight function

$$\rho(x) = \frac{1}{\sqrt{1 - x^2}}$$

as

$$(g_i, g_j) = \int_a^b \rho(x)g_i(x)g_j(x)dx.$$

One may produce the Chebyshev polynomials by executing the Gram–Schmidt procedure. This technique, due to the weight function, is rather tedious. An alternative way to produce them is by the following definition of the kth Chebyshev polynomial:

$$T_k(x) = \cos(k\,arccos(x)); \quad k = 0, 1, \ldots, n.$$

With the substitution of

$$x = cos(\theta),$$

the definition is

$$T_k(x) = cos(k\theta).$$

Well-known simple trigonometric identities of

$$cos((k+1)\theta) = cos(\theta)cos(k\theta) - sin(\theta)sin(k\theta),$$

and

$$cos((k-1)\theta) = cos(\theta)cos(k\theta) + sin(\theta)sin(k\theta)$$

may be combined as

$$cos((k+1)\theta) = 2cos(\theta)cos(k\theta) - cos((k-1)\theta).$$

Substitution yields a three-member recurrence formula for the Chebyshev polynomials.

$$T_{k+1}(x) = 2xT_k(x) - T_{k-1}(x).$$

Starting again with

$$T_0(x) = 1$$

and

$$T_1(x) = x,$$

one proceeds as

$$T_2(x) = 2x^2 - 1$$

and

$$T_3(x) = 4x^3 - 3x.$$

The first four Chebyshev polynomials are shown in Figure 4.2. The Chebyshev polynomials also exhibit interesting characteristics. They are also specially normalized as

$$T_k(1) = 1$$

and

$$T_k(-1) = (-1)^k.$$

Furthermore, they satisfy

$$|T_k(x)| \leq 1; -1 \leq x \leq 1$$

and their symmetry characteristics are similar to those of the Legendre polynomials

$$T_k(-x) = (-)^k T_k(x).$$

The Chebyshev polynomials may also be used to approximate functions similarly to the Legendre polynomials. One can compute coefficients a_k to obtain

$$g(x) = \sum_{k=0}^{n} a_k T_k(x).$$

The most practical usage of the T_k polynomials is, however, in a different form, as discussed in the next section.

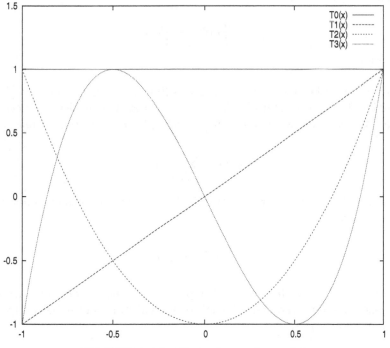

FIGURE 4.2 Chebyshev polynomials

4.3.1   Collapsing a power series

The trigonometric definition of the Chebyshev polynomials enables spectacular possibilities. For example, using the identity

$$x^k = cos^k(\theta) = (\frac{e^{i\theta} + e^{-i\theta}}{2})^k$$

and the binomial theorem, and using the definition of the first two Chebyshev polynomials

$$1 = T_0(x),$$

$$x = T_1(x),$$

and by appropriately grouping like terms, we obtain

$$x^2 = \frac{1}{4}(e^{i2\theta} + 2 + e^{-i2\theta}) = \frac{1}{2}T_0(x) + \frac{1}{2}T_2(x).$$

Similarly,

$$x^3 = \frac{3}{4}T_1(x) + \frac{1}{4}T_3(x),$$

and so forth.

The idea of collapsing a power series was first proposed by Lanczos [8]. The original and very powerful idea is as follows. Let us consider a function for which a power series approximation, for example, a Taylor polynomial, exists.

$$g(x) = \sum_{k=0}^{n} c_k x^k.$$

Substitute x^k powers in terms of the Chebyshev polynomials from above and reorder to produce

$$g(x) = \sum_{k=0}^{n} b_k T_k(x).$$

The concept of collapsing means dropping the highest order term from the Chebyshev-based polynomial. The new series will still result in better accuracy than the power series.

4.3.2   Computational example

Let us consider

$$f(x) = sin(x)$$

in the interval $[-1 \leq x \leq 1]$. If we approximate $f(x)$ by the first three terms of the power series expansion

$$sin(x) = \sum_{k=0}^{\infty}(-1)^k \frac{x^{2k+1}}{(2k+1)!},$$

we obtain the fifth order approximate function of

$$g(x) = x - \frac{1}{6}x^3 + \frac{1}{120}x^5.$$

FIGURE 4.3 Chebyshev example

This has an approximation error of

$$|f(x) - g(x)| < \frac{1}{7!} = \frac{1}{5040}.$$

We employ the binomial theorem again to obtain the x^5 power series member in terms of the Chebyshev polynomials.

$$x^5 = \frac{1}{32}(e^{i5\theta} + 5e^{i3\theta} + 10e^{i\theta} + 10e^{-i\theta} + 5e^{-i3\theta} + e^{-i5\theta}).$$

Substitution yields

$$x^5 = \frac{5}{8}T_1(x) + \frac{5}{16}T_3(x) + \frac{1}{16}T_5(x).$$

We replace the power series terms as

$$g(x) = T_1(x) - \frac{1}{6}\left(\frac{3}{4}T_1(x) + \frac{1}{4}T_3(x)\right) + \frac{1}{120}\frac{1}{16}(10T_1(x) + 5T_3(x) + T_5(x)).$$

Reordering results in

$$g(x) = \frac{169}{192}T_1(x) - \frac{5}{128}T_3(x) + \frac{1}{1920}T_5(x).$$

We drop the last term (we have never computed yet in this book anyway) and obtain

$$g(x) = \frac{169}{192}T_1(x) - \frac{5}{128}T_3(x) = 0.9974x - 0.1562x^3.$$

Figure 4.3 shows the original $sin(x)$ function, the cubic power series $P(x)$ and the Chebyshev-based approximation, noted on the figure as $T(x)$. In the interval of approximation, the Chebyshev approximation is indistinguishable from the $sin(x)$ curve. The power series expansion has a noticeably larger error. The error of the Chebyshev polynomial approximation is less than $\frac{1}{1920}$, which is about a third of the error of the original power series approximation.

4.4   Fourier approximation

The trigonometric foundation of the Chebyshev polynomials hints toward another approximation approach for functions that are periodic in nature. Let us assume that $f(x)$ is periodic with period 2π and integrable in the interval $[-\pi, \pi]$ along with its square. The Fourier, or trigonometric, approximation seeks the following approximation function:

$$g(x) = a_0 + \sum_{k=1}^{n}(a_k cos(kx) + b_k sin(kx)),$$

such that

$$\|f(x) - g(x)\|^2 = min.$$

This form of approximation implies that the basis function set consists of

$$g_0(x) = 1, g_1(x) = cos(x), g_2(x) = sin(x), \ldots,$$

and in general,

$$g_j(x) = \begin{cases} cos(kx), j = 2k - 1, \\ sin(kx), j = 2k. \end{cases}$$

In order to verify that these basis functions are orthogonal, we need to evaluate some integrals for various products of these functions. First the product of two differently indexed ($i \neq j$) even and odd functions is considered.

$$\int_{-\pi}^{\pi} sin(ix)cos(jx)dx = \frac{1}{2}\int_{-\pi}^{\pi} sin((i+j)x) + sin((i-j)x)dx = 0.$$

The substitution of the integrand of the left-hand side is obtained by adding the well-known trigonometric identities of the sin of the sum and difference of two angles. All such integrals are zero since the *sin* function is an odd function and the interval of integration is symmetric with respect to the origin. Similar considerations can be used to prove that

$$\int_{-\pi}^{\pi} cos(ix)cos(jx)dx = 0,$$

and

$$\int_{-\pi}^{\pi} sin(ix)sin(jx)dx = 0.$$

Let us now focus on products with identical indices. For an even term

$$\int_{-\pi}^{\pi} sin(jx) \cdot sin(jx)dx = \int_{-\pi}^{\pi}(\frac{1}{2} - \frac{1}{2}cos(2jx))dx = \pi,$$

and for an odd term

$$\int_{-\pi}^{\pi} cos(jx) \cdot cos(jx)dx = \int_{-\pi}^{\pi}(\frac{1}{2} + \frac{1}{2}cos(2jx))dx = \pi.$$

Thus it follows that the set is orthogonal with respect to the integral

$$(g_i, g_j) = \int_{-\pi}^{\pi} g_i(x)g_j(x)dx = \begin{cases} 0, i \neq j, \\ \pi, i = j. \end{cases}$$

Finally we create a basis function set (G_j) from which the the unknown coefficients are easily computed. Since

$$||g_0||^2 = \int_{-\pi}^{\pi} 1 \cdot 1 dx = 2\pi,$$

the first term of the new set is

$$G_0(x) = \frac{1}{2\pi}.$$

Then

$$||g_1||^2 = \int_{-\pi}^{\pi} cos(x) \cdot cos(x) dx = \pi$$

results in

$$G_1(x) = \frac{1}{\pi} cos(x).$$

Similarly

$$||g_2||^2 = \int_{-\pi}^{\pi} sin(x) \cdot sin(x) dx = \pi$$

yields

$$G_2(x) = \frac{1}{\pi} sin(x).$$

In general for any $j > 0$,

$$||g_j||^2 = \pi,$$

and

$$G_j(x) = \begin{cases} \frac{1}{\pi} cos(kx), j = 2k - 1, \\ \frac{1}{\pi} sin(kx), j = 2k. \end{cases}$$

Then the coefficients may be computed as

$$c_j = (f, G_j), j = 0, 1, \ldots, 2n.$$

The specific first constant is computed as

$$a_0 = \int_{-\pi}^{\pi} f(x) \frac{1}{2\pi} dx,$$

resulting in

$$a_0 = \frac{1}{2\pi} \int_{-\pi}^{\pi} f(x) dx.$$

The odd indexed terms become

$$a_k = \frac{1}{\pi} \int_{-\pi}^{\pi} f(x) cos(kx) dx,$$

and the even indexed terms are

$$b_k = \frac{1}{\pi} \int_{-\pi}^{\pi} f(x) sin(kx) dx.$$

The computation of these coefficients is somewhat tedious as, depending on the type of the function $f(x)$, the integrals may not always be available in closed form.

4.4.1 Computational example

One cannot find a simpler example to demonstrate this process than the approximation of the power function $y = x^2$ in the interval $[-\pi, \pi]$. We seek a

FIGURE 4.4 Fourier example

second order Fourier approximation.

The first coefficient is the product of the integral

$$a_0 = \frac{1}{2\pi} \int_{-\pi}^{\pi} x^2 dx = \frac{1}{3}\pi^2.$$

The generic a coefficients are

$$a_1 = \frac{1}{\pi} \int_{-\pi}^{\pi} x^2 cos(x) dx = -4,$$

and

$$a_2 = \frac{1}{\pi} \int_{-\pi}^{\pi} x^2 cos(2x) dx = 1.$$

The lone b coefficient is

$$b_1 = \frac{1}{\pi} \int_{-\pi}^{\pi} x^2 sin(x) dx = 0.$$

Hence the approximation result is

$$g(x) = \frac{\pi^2}{3} - 4cos(x) + cos(2x).$$

The original function $f(x)$ and the approximation $g(x)$ are shown in Figure 4.4.

4.4.2   Complex Fourier approximation

It is also possible and sometimes advantageous to describe the approximation in complex form as

$$g(x) = \sum_{k=-n}^{n} c_k e^{ikx},$$

where the complex coefficient is

$$c_k = \frac{1}{2\pi} \int_{-\pi}^{\pi} f(x) e^{-ikx} dx.$$

Here $i = \sqrt{-1}$, the imaginary unit, and note the specific index sequencing. The expansion of the approximation is of the form

$$g(x) = c_0 + c_1 e^{ix} + c_{-1} e^{-ix} + \ldots + c_n e^{inx} + c_{-n} e^{-inx}.$$

The complex coefficients are also based on certain orthogonality conditions. Specifically for $k, l = 0, 1, \ldots, n$, it is easy to see that

$$\int_{-\pi}^{\pi} e^{ikx} e^{-ikx} dx = 2\pi,$$

while

$$\int_{-\pi}^{\pi} e^{ikx} e^{ilx} dx = 0, l \neq -k.$$

Multiplying both sides of the approximation function by e^{-ikx} and integrating over the interval of $[-\pi, \pi]$, due to the above orthogonality conditions, there is only one term left standing:

$$\int_{-\pi}^{\pi} f(x) e^{-ikx} dx = \int_{-\pi}^{\pi} c_k e^{ikx} e^{-ikx} dx = 2\pi c_k.$$

This is the equation for the complex coefficients c_k. Note that all the complex coefficients come from a single common formula.

The relationship with the real coefficients may be established by using one of Euler's formulae:

$$e^{-ikx} = cos(kx) - i\, sin(kx).$$

Multiplying by $f(x)$ and integrating yields

$$\int_{-\pi}^{\pi} f(x) e^{-ikx} dx = \int_{-\pi}^{\pi} f(x) cos(kx) dx - i \int_{-\pi}^{\pi} f(x) sin(kx) dx.$$

Substituting the coefficients, we get

$$2\pi c_k = \pi a_k - i\pi b_k,$$

or

$$c_k = \frac{a_k}{2} - i\frac{b_k}{2}.$$

Conversely, since

$$cos(kx) = \frac{e^{ikx} + e^{-ikx}}{2},$$

and

$$sin(kx) = \frac{e^{ikx} - e^{-ikx}}{2i},$$

similar multiplication and integration results in

$$a_k = c_k + c_{-k},$$

and

$$b_k = \frac{c_k - c_{-k}}{i}.$$

Note that the function resulting from the complex Fourier approximation may still be real, even with the complex coefficients and terms.

The complex Fourier approximation technique provides the basis for a clever computational arrangement known as the fast Fourier transform (FFT) [2]. This method has gained wide acceptance in engineering applications and is widely used in software packages.

4.5 Padé approximation

The final technique discussed in this chapter is again of a different class, namely, the approximation function is a rational polynomial. This approximation is executed in the neighborhood of a fixed point. For convenience let us use zero as this point. Then the Padé approximation will be a generalization of the MacLaurin polynomial. Choosing a nonzero fixed point results in a generalization of the Taylor polynomial.

$$g(x) = \frac{p(x)}{q(x)} = \frac{a_0 + a_1 x + a_2 x^2 + \ldots + a_n x^n}{b_0 + b_1 x + b_2 x^2 + \ldots + b_m x^m}.$$

In order to have a nonzero denominator, the condition of

$$b_0 \neq 0$$

is needed. We will enforce this with

$$b_0 = 1.$$

We need to find coefficients a_k and b_k such that the derivatives of the function are approximated as

$$f^{(k)}(0) = g^{(k)}(0), k = 0, 1, \ldots, m + n.$$

When $m = 0$, the Padé approximation is simply the MacLaurin series. The error of the approximation is

$$f(x) - g(x) = f(x) - \frac{p(x)}{q(x)} = \frac{f(x)q(x) - p(x)}{q(x)}.$$

Let us replace the function $f(x)$ to be approximated with its MacLaurin series,

$$f(x) = \sum_{i=0}^{\infty} c_i x^i.$$

Substituting this and the approximating rational function into the error formula gives

$$f(x) - g(x) = \frac{\sum_{i=0}^{\infty} c_i x^i \sum_{i=0}^{m} b_i x^i - \sum_{i=0}^{n} a_i x^i}{q(x)}.$$

Expanding the sums, the numerator is of the form

$$(c_0 + c_1 x + c_2 x^2 + \ldots)(1 + b_1 x + b_2 x^2 + \ldots + b_m x^m) - (a_0 + a_1 x + a_2 x^2 + \ldots + a_n x^n).$$

Note that we took advantage of the enforced b_0 value. Multiplying and re-ordering produces the coefficient of the x^k term as

$$\sum_{i=0}^{k} (c_i b_{k-i}) - a_k.$$

We select the coefficients such that this expression is zero for $k \leq m + n$. This assures that $f(x) - g(x)$ has a zero of multiplicity $m + n + 1$ at $x = 0$. More on this will be discussed in Chapter 7, on the subject of solution of algebraic equations. This results in a set of $m + n + 1$ linear equations of the form

$$\sum_{i=0}^{k} c_i b_{k-i} - a_k = 0, k = 0, 1, \ldots, m + n.$$

This is a homogeneous system of linear equations in $m + n + 1$ unknowns $a_k, k = 0, 1, \ldots, n$ and $b_k, k = 1, \ldots, m$. This process definitely warrants an enlightening example.

4.5.1 Computational example

We consider the approximation of

$$f(x) = e^x.$$

The MacLaurin series expansion is

$$e^x = 1 + x + \frac{1}{2}x^2 + \frac{1}{6}x^3 + \ldots.$$

Let us aim for a Padé approximation with $n = 2$ and $m = 1$, i.e., a quadratic numerator and a linear denominator polynomial.

Following the above theory, the expansion of the numerator is

$$(1 + x + \frac{1}{2}x^2 + \frac{1}{6}x^3 + \ldots)(1 + b_1 x) - (a_0 + a_1 x + a_2 x^2).$$

Since $m + n = 2 + 1 = 3$, we need four equations. Executing the multiplication results in

$$1 - a_0 + (1 + b_1 - a_1)x + (\frac{1}{2} + b_1 - a_2)x^2 + (\frac{1}{6} + \frac{1}{2}b_1)x^3.$$

Collecting out the coefficients for the increasing power terms gives the system of equations as follows:

$$1 - a_0 = 0,$$
$$1 + b_1 - a_1 = 0,$$
$$\frac{1}{2} + b_1 - a_2 = 0,$$

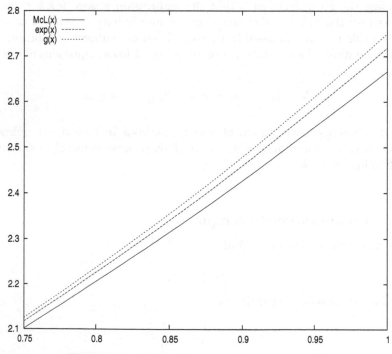

FIGURE 4.5 Padé approximation example

and

$$\frac{1}{6} + \frac{1}{2}b_1 = 0.$$

The first equation gives

$$a_0 = 1,$$

the last

$$b_1 = -\frac{1}{3}.$$

The two intermediate equations produce

$$a_1 = \frac{2}{3},$$

and

$$a_2 = \frac{1}{6}.$$

The Padé approximation of e^x is therefore of the form

$$g(x) = \frac{1 + \frac{2}{3}x + \frac{1}{6}x^2}{1 - \frac{1}{3}x}.$$

Figure 4.5 shows the e^x function, its MacLaurin polynomial as $McL(x)$ and Padé approximations $g(x)$ in the $[0.75, 1]$ interval to be able to visualize the advantage of the Padé approximation. One can verify that

$$g'(0) = g''(0) = g^{(3)}(0) = 0,$$

in agreement with the derivatives of e^x as it was intended. In general, the Padé approximation is superior to the MacLaurin. For example, at $x = 1$ the value of the MacLaurin polynomial is 2.67, while the Padé approximation's is 2.75, the latter clearly closer to $e = 2.7183$.

The Padé approximations have a very wide area of application and the theory is very extensive and deep. Brezinski [1] is the preeminent contemporary researcher of this technique, with dozens of publications.

The topic of the approximation of functions is of course very far from being exhausted. The orthogonal polynomial technique with a weight function of $\rho(x) = e^{-x}$ yields the Laguerre polynomials, and with $\rho(x) = e^{-x^2}$ produces the Hermite polynomials. The details of these are found in advanced approximation texts such as [7] or [11], but as these techniques are not widely used in engineering practice, they are not discussed here further.

4.6   Approximating matrix functions

It is an important case of approximating functions with a matrix as the independent variable. The goal in this case is to find an infinite series for which

$$f(A) = \sum_{k=0}^{\infty} c_k A^k.$$

Let us consider a finite sum of the terms in the series

$$S_n(A) = \sum_{k=0}^{n} c_k A^k$$

and find an appropriate selection of the coefficients. The minimal polynomial of the matrix at hand may be written as

$$D(x) = \prod_{i=1}^{s} (x - \lambda_i)^{m_i}.$$

For simplicity of the discussion we assume that all eigenvalues are simple and $m_i = 1$. Dividing by $D(x)$, we obtain

$$S_n(x) = D(x)Q_n(x) + R_n(x).$$

Since all eigenvalues satisfy the minimal polynomial, $D(\lambda_i) = 0$, it follows that

$$S_n(\lambda_i) = R_n(\lambda_i).$$

Approximating the sum of the remainder series by Lagrange polynomials, introduced in Section 1.2, in the form

$$L_i(x) = \frac{D(x)}{D'(x)(x - \lambda_i)}, i = 1, 2, ..., s,$$

we can compute

$$R_n(x) = \sum_{i=1}^{s} S_n(\lambda_i) L_i(x).$$

Therefore the approximating sum is

$$S_n(A) = R_n(A) = \sum_{i=1}^{s} S_n(\lambda_i) L_i(A).$$

Assuming that all the eigenvalues of the matrix are finite, the approximating series is bounded

$$\lim_{n \to \infty} S_n(\lambda_i) = f(\lambda_i)$$

and the matrix function is approximated by

$$f(A) = \sum_{i=1}^{s} f(\lambda_i)L_i(A).$$

In the case when not all the eigenvalues of the matrix are simple, Hermite polynomials may be used with the multiplicities of the eigenvalues. While the accuracy of this approximation is excellent, the cost of computing the eigenvalues may be prohibitive.

There are other approximation techniques for matrix functions, most notably by using the Padé technique introduced for scalar functions in the previous section, but they will not be discussed in further detail.

4.6.1   Taylor series approximation of matrix functions

To overcome the necessity of computing the eigenvalues, the Taylor series approximation of the corresponding scalar function may be used. For the frequently occurring matrix exponential function, the Taylor series approximation has the form

$$e^A = I + A + \frac{1}{2!}A^2 + ... + \frac{1}{k!}A^k...$$

Since this approximation is also a polynomial, albeit a matrix polynomial, we can use the following reverse scheme to evaluate. Let us precompute the scalar coefficients $c_j = \frac{1}{j!}, j = 0, 1, .., k$. The scheme applied to our case is then as follows:

$S = c_k A + c_{k-1}I$
For $j = k - 2, ..., 0$ compute
$S = AS + c_j I.$

Upon completion of this algorithm $e^A = S$ with the price of $(k-1)n^3$ floating point operations where n is the order of the matrix. This scheme is actually a variation of Horner's method, to be discussed in detail in Section 7.3.2.

The numerical computation of the matrix exponential is fraught with the danger of accumulating computational errors and as such is rather sensitive; see [12]. However, depending on the matrix characteristics and the accuracy requirements, often only the first three or four terms are sufficient in industrial applications. This computation is instrumental in the solution of initial value problems of the type

$$\dot{X}(t) = AX(t)$$

whose solution is

$$X(t) = e^{At}$$

for an initial value of $X(0) = I$. The solution of that problem will be discussed in Section 11.6 in more detail.

Another practical case is the approximation of trigonometric matrix functions. The two most commonly occurring functions are

$$sin(A) = A - \frac{1}{3!}A^3 + \frac{1}{5!}A^5 - +...(-1)^k \frac{1}{(2k+1)!}A^{2k+1}...$$

and

$$cos(A) = I - \frac{1}{2!}A^2 + \frac{1}{4!}A^4 - +...(-1)^k \frac{1}{(2k)!}A^{2k}...$$

Note that these series do not contain all the power terms; the *sin* contains the odd and the *cos* contains the even powers.

Finally, the computation of the logarithm of matrices is an important topic in system models. Let us briefly show a particular approximation:

$$log(I - A) = -A - \frac{1}{2}A^2 - ... - \frac{1}{k}A^k...$$

See [6] for more details.

References

[1] Brezinski, C.; Padé type approximation and general orthogonal polynomials, *Int. Ser. Num. Math.*, Vol. 50, Birkhauser, Basel, 1980.

[2] Cooley, J. W. and Tukey, J. W.; An algorithm for the machine calculation of complex Fourier series, *Mathematics of Computation*, Vol. 19, No. 90, pp. 297–301, 1965.

[3] Chebyshev, P. L.; Sur les questions de minima gui se rattachent a la representation approximative des fonctions, *Mem. Acad. Imp. Sci.*, Vol. 6, 1859.

[4] Franklin, Ph.; *Fourier Methods*, McGraw-Hill, New York, 1949.

[5] Gram, J. P.; Über the Entwicklung reeller Funktionen in Reihen mittels der Methode der kleinsten Quadrate, *J. Reine Angew. Math.*, Vol. 94, pp. 41–73, 1883.

[6] Helton, B. W.; Logarithms of matrices, *Proc. American Mathematical Society*, Vol 19, pp 733–736, 1968.

[7] Hildebrand, F. B.; *Introduction to Numerical Analysis*, MacGraw-Hill, New York, 1970.

[8] Lanczos, C.; *Applied Analysis*, Prentice-Hall, 1956.

[9] Legendre, A, M.; *Nouvelles methodes pour la determination des orbites des cometes*, Courcier, Paris, 1806.

[10] Padé, H.; Sur la generalization des fractions continues algebriques, *J. Math. Pures Appl.*, Vol. IV/10, pp. 291–329, 1894.

[11] Scheid, F.; *Numerical Analysis*, McGraw-Hill, New York, 1968.

[12] Van Loan, C. F.; The sensitivity of the matrix exponential, *SIAM Journal of Numerical Analysis*, Vol 14, pp. 971–981, 1977.

5

Numerical differentiation

As in the last chapter, the input datum of our consideration is still a function. Here, however, we will extract a single quantity related to the function, namely, its derivative. The topic of calculating the derivative of a function approximately (numerical differentiation) has been the focus of interest of mathematicians for many centuries. As will be seen, most of the methods are some kind of an extension of polynomial approximations, such as Lagrange's or Taylor's.

5.1   Finite difference formulae

Based on the teaching of calculus, the definition of the derivative of a continuous function $f(x)$ at the point $x - x_0$ is

$$f'(x_0) = \lim_{h \to 0} \frac{f(x_0 + h) - f(x_0)}{h}.$$

For a small, finite h value the approximation of

$$f'(x_0) \approx \frac{f(x_0 + h) - f(x_0)}{h}$$

is commonly applied, that is, of course, approximating the slope of the tangent of the function (the derivative) with the slope of a chord in the neighborhood of the point of interest. In other words, the differential is approximated by a finite difference, hence the name of this class of methods.

This is a conceptually very simple method, but we would like to also have an error estimate. To do so, we utilize the Lagrange polynomial approximation of Chapter 1. Introduce a point

$$x_1 = x_0 + h$$

and generate a first order Lagrange polynomial through the two points. Based on the discussion in Chapter 1, one can write

$$f(x) = \sum_{k=0}^{1} f(x_k)L_k(x) + \frac{(x - x_0)(x - x_1)}{2!}f''(\xi),$$

where the last term is the error formula of the Lagrange approximation and

$$x_0 < \xi < x_1.$$

Substituting the Lagrange base polynomials tailored for our case,

$$f(x) = f(x_0)\frac{x - x_1}{x_0 - x_1} + f(x_1)\frac{x - x_0}{x_1 - x_0} + \frac{(x - x_0)(x - x_1)}{2}f''(\xi).$$

Furthermore,

$$f(x) = f(x_0)\frac{x - x_0 - h}{-h} + f(x_0 + h)\frac{x - x_0}{h} + \frac{(x - x_0)(x - x_0 - h)}{2}f''(\xi).$$

Analytically differentiating and substituting yields

$$f'(x_0) = \frac{f(x_0 + h) - f(x_0)}{h} - \frac{h}{2}f''(\xi).$$

It is important to point out that, despite the definite negative sign of the error term, the correction will be in the appropriate direction. That is assured by the fact that the second derivative is part of the error term. If the second derivative is negative, the curve is convex from downward, and the chord-based slope is an underestimate. The error term is positive in this case and vice versa. Unfortunately, the error term cannot be computed precisely, as it is taken at an unknown location.

5.1.1   Three-point finite difference formulae

The Lagrange polynomial based approach, besides producing a valuable error term, permits a useful generalization. Let us consider $n + 1$ points, one of which is the point of interest,

$$x_k, k = 0, 1, \ldots, n.$$

Following the process introduced above,

$$f(x) = \sum_{k=0}^{n} f(x_k)L_k(x) + \frac{(x - x_0)\cdots(x - x_n)}{(n+1)!}f^{n+1}(\xi),$$

where

$$x_0 < \xi < x_n.$$

Differentiating again and substituting the point of interest, say, x_j, we obtain a general $n + 1$ point difference formula.

$$f'(x_j) = \sum_{k=0}^{n} f(x_k) L_k'(x_j) + \prod_{k=0, k \neq j}^{n} (x_j - x_k) \frac{f^{(n+1)}(\xi)}{(n+1)!}.$$

Depending on the choice of the point of interest, we obtain the above forward finite difference formula when $x_j = x_0$, a backward finite difference formula when $x_j = x_n$, and a centered finite difference formula otherwise.

In engineering practice, the 3-point centered finite difference formula is widely used and will be discussed here. For this case, the first Lagrange base polynomial and its derivative are

$$L_0(x) = \frac{(x - x_1)(x - x_2)}{(x_0 - x_1)(x_0 - x_2)},$$

and

$$L_0'(x) = \frac{2x - x_1 - x_2}{(x_0 - x_1)(x_0 - x_2)}.$$

Similarly the second and third Lagrange basis polynomials' first derivatives are

$$L_1'(x) = \frac{2x - x_0 - x_2}{(x_1 - x_0)(x_1 - x_2)},$$

and

$$L_2'(x) = \frac{2x - x_0 - x_1}{(x_2 - x_0)(x_2 - x_1)}.$$

With the above terms, the generic 3-point approximate derivative formula is

$$f'(x_j) = f(x_0) \frac{2x_j - x_1 - x_2}{(x_0 - x_1)(x_0 - x_2)} + f(x_1) \frac{2x_j - x_0 - x_2}{(x_1 - x_0)(x_1 - x_2)}$$

$$+ f(x_2) \frac{2x_j - x_0 - x_1}{(x_2 - x_0)(x_2 - x_1)} + \prod_{k=0, k \neq j}^{2} (x_j - x_k) \frac{f^{(3)}(\xi)}{6}.$$

In engineering practice the points sampled or measured are quite often equidistant. We have already taken advantage of this on prior occasions and will also do so here. We specify

$$x_j - x_{j-1} = h.$$

Assuming that $x_j = x_0$ and

$$x_1 = x_0 - h,$$

and

$$x_2 = x_0 + h$$

produces the equation

$$f'(x_j) = f'(x_0) = f(x_0)\frac{2x_0 - (x_0 - h) - (x_0 + h)}{h(-h)}$$

$$+ f(x_0 - h)\frac{2x_0 - x_0 - (x_0 + h)}{-h(-2h)} + f(x_0 + h)\frac{2x_0 - x_0 - (x_0 - h)}{h(2h)}$$

$$+ \frac{-h^2}{6}f^{(3)}(\xi).$$

Sorting results in the 3-point, equidistant centered difference formula of

$$f'(x_0) = \frac{f(x_0 + h) - f(x_0 - h)}{2h} - \frac{h^2}{6}f^{(3)}(\xi).$$

Here

$$x_0 - h < \xi < x_0 + h.$$

Note that this 3-point centered difference formula has an error of order h^2 as opposed to the earlier order h. At the boundary of the interval of interest (or the continuity of the function), one may desire to again set $x_j = x_0$, but with

$$x_1 = x_0 + h,$$

and

$$x_2 = x_0 + 2h.$$

Then the formula becomes a 3-point forward difference formula of

$$f'(x_0) = \frac{-3f(x_0) + 4f(x_0 + h) - f(x_0 + 2h)}{2h} + \frac{h^2}{3}f^{(3)}(\xi).$$

Here

$$x_0 < \xi < x_0 + 2h.$$

An interesting application of the latter formula is to approximate end slopes for splines. Changing h to $-h$ results in a 3-point backward difference formula. Note that the error of these formulae is still $O(h^2)$, due to their centered difference foundation, as opposed to the original forward or backward difference formula's error of $O(h)$.

From the error terms introduced so far it seems like reducing the size of h is advantageous. One must exercise caution in this regard, however, as the smaller the h gets the closer the terms in the numerator get. As subtraction of very close numbers in finite precision arithmetic is numerically dangerous, clearly reducing h to a very small value holds its own pitfall.

This avenue is not explored here further; the remarks are meant to point out the inherently unstable nature of numerical differentiation. This is the area where the distinction mentioned in the introduction about formula error and numerical computation error is relevant. Modern numerical analysis texts such as [4] deal with the detailed analysis of the latter class of errors.

5.1.2 Computational example

Let us consider finding the derivative of $f(x) = sin(x)$ using the points of

$$x_0 = 0.5, x_1 = 0.6, x_2 = 0.7.$$

This is a computational example where the use of a calculator or a computer program is required to evaluate the function at the desired points. The appropriate function values are

$$f(x_0) = sin(0.5) = 0.4794,$$
$$f(x_1) = sin(0.6) = 0.5646,$$
$$f(x_2) = sin(0.7) = 0.6442.$$

The calculation at the center point is simply the application of the 3-point centered difference formula as

$$f'(x_1) = f'(0.6) = \frac{sin(0.7) - sin(0.5)}{0.2} = 0.8239.$$

The error term for this case is bounded by

$$\frac{0.1^2}{6}cos(\xi) \le \frac{0.1^2}{6}cos(x_0) = 0.0014.$$

Here we have chosen $x_0 = \xi$ to maximize the absolute value of the error term in the interval. This fits well with the theoretical value of

$$sin'(x)|_{x=0.6} = cos(0.6) = 0.8253.$$

Figure 5.1 shows the function and its numerical derivative-based tangent line as t1(x). In order to compute the derivative at the left end point, the 3-point forward difference formula is used:

$$f'(x_0) = f'(0.5) = \frac{-3sin(0.5) + 4sin(0.6) - sin(0.7)}{0.2} = 0.8804.$$

The error term is

$$|-\frac{0.1^2}{3}cos(\xi)| \le |-\frac{0.1^2}{3}cos(x_0)| = |-0.0028|.$$

The exact value is

$$sin'(0.5) = cos(0.5) = 0.8776.$$

Finally, the right end point derivative may be computed from the 3-point backward difference formula by replacing h with $-h$ as

$$f'(x_2) = f'(0.7) = \frac{sin(0.5) - 4sin(0.6) + 3sin(0.7)}{0.2} = 0.7676,$$

which also compares well with the exact value of

$$sin'(0.7) = cos(0.7) = 0.7648.$$

5.2   Higher order derivatives

Until now we have focused on the first derivative only. We can also compute higher order derivatives by utilizing a sequence of forward differences. With

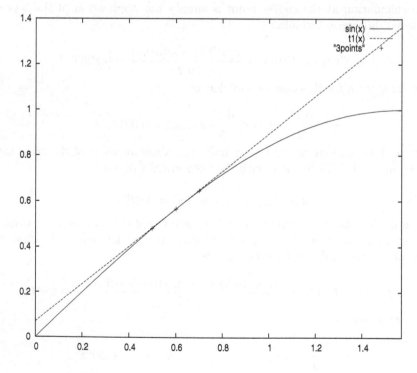

FIGURE 5.1 Numerical differentiation example

$$\Delta f(x_0) = f(x_0 + h) - f(x_0),$$

the forward difference based first derivative derived in Section 5.1 is

$$f'(x_0) = \frac{1}{h} \Delta f(x_0).$$

Introducing the second order forward difference

$$\Delta^2 f(x_0) = \Delta f(x_0 + h) - \Delta f(x_0) = f(x_0 + 2h) - 2f(x_0 + h) + f(x_0),$$

the approximate second derivative is written as

$$f''(x_0) = \frac{\Delta^2 f(x_0)}{h^2}.$$

The obvious path of generalization is by introducing

$$\Delta^k f(x_0) = \Delta^{k-1} f(x_0 + h) - \Delta^{k-1} f(x_0).$$

Then any higher order derivative is obtained by

$$f(k)(x_0) = \frac{\Delta^k f(x_0)}{h^k}.$$

Executing the differences recursively, for $k = 3$ we obtain

$$f'''(x_0) = \frac{\Delta^k f(x_0)}{h^k} = \frac{f(x_0 + 3h) - 3f(x_0 + 2h) + 3f(x_0 + h) - f(x_0)}{h^3}.$$

We observe that the coefficients are the binomial coefficients, hence

$$\Delta^k f(x_0) = \sum_{i=0}^{k} (-1)^k \binom{k}{i} f(x_0 + (k - i)h).$$

For example, the fourth order forward difference is

$$\Delta^4 f(x_0) = (-1)^0 \binom{4}{0} f(x_0 + 4h) + (-1)^1 \binom{4}{1} f(x_0 + 3h)$$

$$+ (-1)^2 \binom{4}{2} f(x_0 + 2h) + (-1)^3 \binom{4}{3} f(x_0 + 1h) + (-1)^4 \binom{4}{4} f(x_0 + 0h).$$

Hence the approximate fourth derivative is computed as

$$f^{(4)}(x_0) = \frac{1}{h^4} (f(x_0 + 4h) - 4f(x_0 + 3h) + 6f(x_0 + 2h) - 4f(x_0 + h) + f(x_0)).$$

This offers a more expedient way of generating higher order approximate derivative formulae than the recursive computation from the lower order differences. Note, however, that all these formulae have $O(h)$ error, which may not be adequate. The following section will introduce two techniques to improve the accuracy.

5.3   Richardson extrapolation

The technique of this section aims to improve the accuracy of the low order approximation formulae. The underlying concept [3] is to extrapolate in such a way as to eliminate some of the error, as shown below. Let us consider the Taylor series expansion:

$$f(x_0 + h) = f(x_0) + f'(x_0)h + \frac{1}{2}f''(x_0)h^2 + \frac{1}{6}f^{(3)}(x_0)h^3$$

$$+ \frac{1}{24}f^{(4)}(x_0)h^4 + \frac{1}{120}f^{(5)}(\xi_r)h^5.$$

Similarly toward the left-hand side of the point of interest:

$$f(x_0 - h) = f(x_0) - f'(x_0)h + \frac{1}{2}f''(x_0)h^2 - \frac{1}{6}f^{(3)}(x_0)h^3$$

$$+ \frac{1}{24}f^{(4)}(x_0)h^4 - \frac{1}{120}f^{(5)}(\xi_l)h^5.$$

In the above:

$$x_0 < \xi_r < x_0 + h,$$

and

$$x_0 - h < \xi_l < x_0.$$

Subtracting the two equations and expressing the first derivative yields

$$f'(x_0) = \frac{f(x_0 + h) - f(x_0 - h)}{2h} - \frac{h^2}{6}f^{(3)}(x_0) - \frac{h^4}{120}f^{(5)}(\xi).$$

Here now ξ is

$$x_0 - h < \xi < x_0 + h,$$

and may be found based on the intermediate value theorem. The first part of this formula is the same as the earlier derived 3-point centered difference formula; however, we now have two error terms. The first one is possible to compute precisely as it is taken at the point of interest and therefore may be used as an actual correction term. The second term is still an approximation term taken at an unknown location $x_0 - h < \xi < x_0 + h$.

We will now write this in terms of a different step size of $h/2$ and with the correction term,

$$f'(x_0) = \frac{f(x_0 + h/2) - f(x_0 - h/2)}{h} - \frac{h^2}{24}f^{(3)}(x_0) - \frac{h^4}{1920}f^{(5)}(\xi_2).$$

Here

$$x_0 - h/2 < \xi_2 < x_0 + h/2.$$

Introducing the notation

$$R_1(h) = \frac{f(x_0 + h) - f(x_0 - h)}{2h},$$

and

$$R_1(h/2) = \frac{f(x_0 + h/2) - f(x_0 - h/2)}{h},$$

we write

$$f'(x_0) = R_1(h) - \frac{h^2}{6} f^{(3)}(x_0) - \frac{h^4}{120} f^{(5)}(\xi),$$

and

$$f'(x_0) = R_1(h/2) - \frac{h^2}{24} f^{(3)}(x_0) - \frac{h^4}{1920} f^{(5)}(\xi_2).$$

Multiplying the last equation by 4 and subtracting the previous equation results in

$$3f'(x_0) = 4R_1(h/2) - R_1(h) + \frac{h^4}{480} k,$$

where

$$k = 4f^{(5)}(\xi) + f^{(5)}(\xi_2).$$

The obvious observation and the salient feature of the technique is that the computable error term of $O(h^2)$ has been eliminated. Reordering yields the formula of

$$f'(x_0) = R_1(h/2) + \frac{R_1(h/2) - R_1(h)}{3} + O(h^4),$$

with an improved formula error.

To summarize the concept, Richardson extrapolation improved the accuracy of a certain approximation formula by evaluating it at two different step sizes. Appropriately combining the results produces a much smaller formula error. Note that the concept is independent of the formula used and it may also be applied to other formulae.

Let us introduce the notation

$$R_2(h) = R_1(h/2) + \frac{R_1(h/2) - R_1(h)}{3}.$$

Hence the approximate derivative is

$$f'(x_0) = R_2(h) + O(h^4).$$

Then the Richardson extrapolation may be generalized as follows.

$$R_j(h) = R_{j-1}(h/2) + \frac{R_{j-1}(h/2) - R_{j-1}(h)}{4^{j-1} - 1}, j = 2, 3, \ldots.$$

TABLE 5.1
Richardson extrapolation scheme

Step	$j = 1$	$j = 2$	$j = 3$	$j = 4$
h	$R_1(h)$			
h/2	$R_1(h/2)$	$R_2(h)$		
h/4	$R_1(h/4)$	$R_2(h/2)$	$R_3(h)$	
h/8	$R_1(h/8)$	$R_2(h/4)$	$R_3(h/2)$	$R_4(h)$
Error:	$O(h^2)$	$O(h^4)$	$O(h^6)$	$O(h^8)$

If a first order formula has an error term of $O(h^2)$, then the jth order formula will have an error term of $O(h^{2j})$. To use the higher order Richardson terms, more and more subdivision of the step size is required. Table 5.1 shows the tabulation of the process.

5.3.1   Computational example

Let us consider the example of $f(x) = ln(x)$ and find the derivative at the point $x_0 = 1$. As it is well known analytically, the derivative has a unit value at that point. We will use the centered 3-point formula with Richardson extrapolation and use an initial step size of $h = 0.4$.

The centered 3-point formula is computed as

$$f'(x_0) = \frac{ln(1.4) - ln(0.6)}{2 \cdot 0.4} = 1.0591223.$$

TABLE 5.2
Richardson extrapolation example

h	$R_1(h)$	$R_2(h)$	$R_3(h)$
0.4	1.0591223		
0.2	1.0136628	0.9985096	
0.1	1.0033535	0.9999170	1.0000109

The error is clearly of $O(h^2)$. The Richardson extrapolation process for $j = 2, 3$ is executed in Table 5.2 and clearly demonstrates the method's accuracy advantages.

5.4 Multipoint finite difference formulae

Going from the 2-point forward difference formula to the 3-point formula showed an order of magnitude improvement in error. It seems like the higher the number of points the better the approximation is. The clever scheme for general multipoint techniques developed in [1] and discussed in this section exploits this tendency.

We will assume an equidistant sampling of the function with step size h:

$$x_i = x_0 + ih, i =, 0, 1, \ldots, 2m - 1.$$

We seek the approximate first derivative of $f(x)$ in the form of

$$f'(x_i) = \sum_{k=-m}^{m} c_k f(x_{i+k}).$$

Note the centered difference flavor of the formulation; the point of interest is an internal x_i point and m points to both sides of it are used. Let us further approximate the function values with their Taylor series as

$$f(x_{i+k}) = f(x_i) + khf'(x_i) + \frac{k^2 h^2}{2!} f''(x_i) + \frac{k^3 h^3}{3!} f^{(3)}(x_i) + \frac{k^4 h^4}{4!} f^{(4)}(x_i) + \ldots.$$

Substituting the latter into the approximate derivative form and ordering terms results in

$$f'(x_i) = f(x_i) \sum_{k=-m}^{m} c_k + h f'(x_i) \sum_{k=-m}^{m} k c_k + \frac{h^2}{2!} f''(x_i) \sum_{k=-m}^{m} k^2 c_k +$$

$$+ \frac{h^3}{3!} f^{(3)}(x_i) \sum_{k=-m}^{m} k^3 c_k + \frac{h^4}{4!} f^{(4)}(x_i) \sum_{k=-m}^{m} k^4 c_k + \ldots.$$

Comparing the two sides yields the following $2m + 1$ equations:

$$\sum_{k=-m}^{m} c_k = 0,$$

$$\sum_{k=-m}^{m} kc_k = \frac{1}{h},$$

$$\sum_{k=-m}^{m} k^2 c_k = 0,$$

and so on until

$$\sum_{k=-m}^{m} k^{2m} c_k = 0.$$

This nonhomogeneous system of equations uniquely defines the c_k coefficients of the approximation. Let us consider the case of $m = 2$ and the matrix form

$$A\underline{c} = \underline{b},$$

which is detailed as

$$\begin{bmatrix} 1 & 1 & 1 & 1 & 1 \\ -2 & -1 & 0 & 1 & 2 \\ 4 & 1 & 0 & 1 & 4 \\ -8 & -1 & 0 & 1 & 8 \\ 16 & 1 & 0 & 1 & 16 \end{bmatrix} \begin{bmatrix} c_1 \\ c_2 \\ c_3 \\ c_4 \\ c_{2m+1} \end{bmatrix} = \begin{bmatrix} 0 \\ \frac{1}{h} \\ 0 \\ 0 \\ 0 \end{bmatrix}.$$

The system matrix is of Vandermonde type and thus always invertible. The solution for the specific case yields

$$\underline{c} = \begin{bmatrix} \frac{1}{12h} \\ \frac{-8}{12h} \\ 0 \\ \frac{8}{12h} \\ \frac{-1}{12h} \end{bmatrix}.$$

Since $2m + 1 = 5$, this is a 5-point formula:

$$f'(x_i) = \frac{f(x_{i-2}) - 8f(x_{i-1}) + 8f(x_{i+1}) - f(x_{i+2})}{12h}.$$

Note that the middle coefficient became zero; hence the formula appears to use only 4 points.

The error of the multipoint formulae is proportional to

$$\frac{k^{2m+l} h^{2m+l}}{(2m+l)!} f^{(2m+l)}(\xi),$$

where $2m + l$ is the first nonzero omitted term of the Taylor series, most of the time $l = 1$. The actual error of a certain formula is

$$O(\frac{h^{2m+l}}{h^p}),$$

where p is the power of the denominator term of the formula. The error for the above 5-point form is

$$O(\frac{h^{2\cdot2+1}}{h}),$$

specifically it is $O(h^4)$. A similar computation for $m = 3$ would result in the following 7-point formula of

$$f'(x_i) = \frac{f(x_{i-3}) + 9f(x_{i-2}) - 45f(x_{i-1}) + 45f(x_{i+1}) - 9f(x_{i+2}) + f(x_{i+3})}{60h},$$

with an error term of $O(h^6)$. The middle term is missing again as before.

The procedure simply generalizes to higher order derivatives; the only difference is that the $\underline{b}$ vector has its nonzero term in the location corresponding to the order of the derivative. For the nth derivative it would move into the n location and become

$$\frac{n!}{h^n}.$$

The following multipoint higher order derivatives were obtained by this process and are useful in engineering practice:

A 5-point second derivative,

$$f''(x_i) = \frac{-f(x_{i-2}) + 16f(x_{i-1}) - 30f(x_i) + 16f(x_{i+1}) - f(x_{i+2})}{12h^2}.$$

A 7-point third derivative,

$$f^{(3)}(x_i) = \frac{f(x_{i-3}) - 8f(x_{i-2}) + 13f(x_{i-1}) - 13f(x_{i+1}) + 8f(x_{i+2}) - f(x_{i+3})}{8h^3}.$$

A 7-point fourth derivative,

$$f^{(4)}(x_i) = \frac{-f(x_{i-3}) + 12f(x_{i-2}) - 39f(x_{i-1}) + 56f(x_i)}{6h^4}$$
$$+ \frac{-39f(x_{i+1}) + 12f(x_{i+2}) - f(x_{i+3})}{6h^4}.$$

The above three methods all have an error of $O(h^4)$. Note the occasional appearance of the middle term (the point of interest) as well as the different sign patterns. There is a wealth of multipoint formulae presented in handbooks such as [2].

It is easy to see that using $m = 1$ in this derivation scheme for the first derivative yields

$$f'(x_i) = \frac{-f(x_{i-1}) + f(x_{i+1})}{2h},$$

which is the same as the centered difference formulae already obtained by means of the Lagrange basis functions in Section 5.1.1. Similarly for the higher order derivatives, this process results in the same formulae as the Lagrange-based formulae; however, the Lagrange method is difficult with a large number of points.

Furthermore, the forward and backward difference formulae may also be derived from this multipoint technique. For example, the 5-point forward difference first derivative formula may be obtained by the simple restructuring of the linear system as

$$\begin{bmatrix} 1 & 1 & 1 & 1 & 1 \\ 0 & 1 & 2 & 3 & 4 \\ 0 & 1 & 4 & 9 & 16 \\ 0 & 1 & 8 & 27 & 64 \\ 0 & 1 & 16 & 81 & 256 \end{bmatrix} \begin{bmatrix} c_1 \\ c_2 \\ c_3 \\ c_4 \\ c_{2m+1} \end{bmatrix} = \begin{bmatrix} 0 \\ \frac{1}{h} \\ 0 \\ 0 \\ 0 \end{bmatrix}.$$

The solution of this system is

$$\underline{c} = \begin{bmatrix} \frac{-25}{12h} \\[2mm] \frac{48}{12h} \\[2mm] \frac{-36}{12h} \\[2mm] \frac{16}{12h} \\[2mm] \frac{-3}{12h} \end{bmatrix}.$$

Notice the reappearance of the fifth term. The 5-point forward difference formula is now

$$f'(x_i) = \frac{-25f(x_i) + 48f(x_{i+1}) - 36f(x_{i+2}) + 16f(x_{i+3}) - 3f(x_{i+4})}{12h}.$$

Again substituting $h = -h$ results in the corresponding 5-point backward difference formula of

$$f'(x_i) = \frac{25f(x_i) - 48f(x_{i-1}) + 36f(x_{i-2}) - 16f(x_{i-3}) + 3f(x_{i-4})}{12h}.$$

Finally, it is noteworthy that the multipoint techniques are the basis of the finite difference method for the solution of initial value problems discussed in Chapter 11.

References

[1] Bajcsay, P.; *Numerikus analizis*, Tankönyvkiadó, Budapest, 1972.

[2] Bronshtein, I. N. and Semendyayev, K. A.; *Handbook of Mathematics*, Van Nostrand, New York, 1985.

[3] Richardson, L. F. and Gaunt, J. A.; The deferred approach to the limit, *Phil. Trans. of Royal Society*, Vol. 226A, pp. 299–361, London, 1927.

[4] Hämmerlin, G. and Hoffmann, K.-H.; *Numerical Mathematics*, Springer Verlag, New York, 1991.

6

Numerical integration

The techniques of this chapter are also rather old; the beginning of the 18th century brought the renaissance of this topic, starting with Newton [9]. His work was generalized by Cotes [4], resulting in the class of Newton–Cotes formulae, the subject of the first section of this chapter. Simpson [11], who is credited with giving the trigonometric functions their names, is the source of one of the most widely used formulae discussed below. Stirling [12] is worthy of mention here for suggesting the idea of composite methods. Gauss took up the topic about a century later, and as in many occasions, standing on the shoulders of giants, he added a significant contribution [5].

The input considered in this chapter is still a function. The subject of the approximation is the definite integral of the function in a given interval. As the geometric meaning (in the case of a function of one variable) is the area under the curve, these computations are often called numerical quadrature. The reason for such a calculation may be that the integral is not easy, or even possible, to integrate analytically. There are also other reasons in engineering applications, specifically those arising in finite element analysis.

6.1 The Newton–Cotes class

The Newton–Cotes class of quadrature methods is based on the equidistant subdivision of the interval of integration into n subintervals spanned by $n+1$ points. Hence n is the order of the formula. The first formula we discuss is used very often in engineering applications as it is very easy to compute. It is commonly called the trapezoid formula and it is the $n = 1$ case of the Newton–Cotes formulae.

The approximate computation of

$$I = \int_a^b f(x)dx$$

may be achieved following similar concepts as earlier. For example, the function may be replaced by its Lagrange polynomial, since the polynomials are always integrable. Let us replace the function with its first order Lagrange polynomial and the associated error term:

$$f(x) = f(x_0)L_0(x) + f(x_1)L_1(x) + \frac{1}{2}f''(\xi)(x - x_0)(x - x_1).$$

6.1.1 The trapezoid rule

Assigning first

$$a = x_0, \ b = x_1,$$

expanding the Lagrange basis polynomials and substituting into the integral, we obtain

$$I = \int_{x_0}^{x_1} (f(x_0)\frac{x - x_1}{x_0 - x_1} + f(x_1)\frac{x - x_0}{x_1 - x_0})dx + \int_{x_0}^{x_1} \frac{1}{2}f''(\xi)(x - x_0)(x - x_1)dx.$$

Integration results in

$$I = [\frac{(x - x_1)^2}{2(x_0 - x_1)}f(x_0) + \frac{(x - x_0)^2}{2(x_1 - x_0)}f(x_1)]_{x_0}^{x_1} + \frac{f''(\xi)}{2}[\frac{x^3}{3} - \frac{x_1 + x_0}{2}x^2 + x_0x_1x]_{x_0}^{x_1}.$$

Executing the posed algebraic operations and introducing

$$h = x_1 - x_0$$

yields

$$I = \frac{h}{2}(f(x_0) + f(x_1)) - \frac{h^3}{12}f''(\xi).$$

This is a familiar formula, known as the trapezoid rule. It is easy to see that the first term is the area of the trapezoid bounded by

$$x = x_0, x = x_1, y = 0$$

and the chord going through the points

$$(x_0, f(x_0)), (x_1, f(x_1)).$$

The area under the curve $f(x)$ is approximated by the area of the trapezoid inscribed into the function.

6.1.2 Simpson's rule

Let us now focus on the $n = 2$ case. This means we will use 3 points and we assign the boundary points of the integral as

$$a = x_0$$

and

$$b = x_n = x_2.$$

The intermediate point for this case will be

$$x_1 = x_0 + h$$

with

$$h = \frac{b-a}{n} = \frac{b-a}{2}.$$

Notice the use of the original integral boundaries. The use of the Lagrange basis polynomials for 3 points yields

$$I = \int_{x_0}^{x_1} [f(x_0)L_0(x) + f(x_1)L_1(x) + f(x_2)L_2(x)]dx.$$

Detailing the Lagrange basis polynomials, the integral becomes

$$I = \int_{x_0}^{x_1} [f(x_0)\frac{(x-x_1)(x-x_2)}{(x_0-x_1)(x_0-x_2)} + f(x_1)\frac{(x-x_0)(x-x_2)}{(x_1-x_0)(x_1-x_2)}$$

$$+ f(x_2)\frac{(x-x_0)(x-x_1)}{(x_2-x_0)(x_2-x_1)}]dx + \int_{x_0}^{x_1} f^{(3)}(\xi)\frac{(x-x_0)(x-x_1)(x-x_2)}{6}dx.$$

Integration, substitution and some algebra yield Simpson's rule,

$$\int_a^b f(x)dx = \frac{h}{3}[f(x_0) + 4f(x_1) + f(x_2)] + e.$$

Here e is the error of the formula, which we will temporarily leave as is. One can observe that the formula is accurate for third degree polynomials:

$$\int_a^b x^3 dx = \frac{b-a}{6}(a^3 + 4(\frac{a+b}{2})^3 + b^3) = \frac{b^4 - a^4}{4}.$$

This is clearly equivalent to the result of Simpson's rule,

$$\int_a^b f(x)dx = \frac{b-a}{6}(f(a) + 4f(\frac{a+b}{2}) + f(b)),$$

when

$$f(x) = x^3.$$

The error
$$e = \int_{x_0}^{x_1} f^{(3)}(\xi) \frac{(x - x_0)(x - x_1)(x - x_2)}{6} dx$$
is difficult to compute directly due to the sign changes in the integration interval. Instead, we will derive the error term from the just learned fact that Simpson's rule is exact for up to third order polynomials. Simpson's rule may be written in the form of 3 unknown coefficients:

$$\int_a^b f(x)dx = c_0 f(x_0) + c_1 f(x_1) + c_2 f(x_2).$$

Specifically for $f(x) = x, x^2, x^3$ the exact integrals with $h = (b - a)/2$ are

$$\int_a^b x\,dx = 2hx_0 + 2h^2,$$

$$\int_a^b x^2\,dx = 2hx_0^2 + 4h^2 x_0 + \frac{8h^3}{3},$$

and

$$\int_a^b x^3\,dx = 2hx_0^3 + 6h^2 x_0^2 + 8h^3 x_0 + 4h^4.$$

These are respectively equivalent to

$$c_0 x_0 + c_1(x_0 + h) + c_2(x_0 + 2h),$$
$$c_0 x_0^2 + c_1(x_0 + h)^2 + c_2(x_0 + 2h)^2,$$

and

$$c_0 x_0^3 + c_1(x_0 + h)^3 + c_2(x_0 + 2h)^3.$$

The resulting system of equations has the solution of

$$c_0 = c_2 = \frac{h}{3}$$

and

$$c_1 = \frac{4h}{3}.$$

Now we execute the same procedure again for a fourth order polynomial, for which the formula has an error. The exact integral is

$$\int_{a=x_0}^{b=x_2} x^4\,dx = \frac{1}{5}(x_2^5 - x_0^5).$$

The approximate integral via Simpson's rule is

$$\frac{h}{3}(x_0^4 + 4x_1^4 + x_2^4).$$

Their difference is

$$-\frac{h^5}{90}24,$$

where

$$24 = (x^4)^{(4)} = f^{(4)}(\xi).$$

Hence the error term of Simpson's rule is

$$e = -\frac{h^5}{90}f^{(4)}(\xi).$$

The above result generalizes as follows: the nth order Newton–Cotes formulae with even n are accurate to a polynomial of order $n + 1$, while the odd formulae are accurate only to the same order. The trapezoid rule is odd $(n = 1)$, and exact only for a linear polynomial, while the one order higher $(n = 2)$ Simpson's rule is even, and exact for a cubic polynomial.

The generic error terms following [6] are

$$e_{n=even} = \frac{h^{n+3}f^{(n+2)}(\xi)}{(n+2)!}\int_0^n m^2(m-1)\cdots(m-n)dm$$

and

$$e_{n=odd} = \frac{h^{n+2}f^{(n+1)}(\xi)}{(n+1)!}\int_0^n m^2(m-1)\cdots(m-n)dm.$$

Based on the above, the following Newton's rule, despite being $n = 3$, is exact also only to cubic polynomials.

Let us consider the generic odd error term of

$$e_{n=3} = \frac{h^5 f^{(4)}(\xi)}{24}\int_0^3 m^2(m-1)(m-2)(m-3)dm.$$

Executing the integration yields

$$e_{n=3} = \frac{h^5 f^{(4)}(\xi)}{24}\frac{-9}{10} = -\frac{3h^5 f^{(4)}(\xi)}{80}.$$

Creating the computational part as in earlier cases results in

$$\int_a^b f(x)dx = \frac{3h}{8}(f(x_0) + 3f(x_1) + 3f(x_2) + f(x_3)) + e.$$

Finally the $n = 3$ Newton–Cotes class formula, also known for its notable constant as Newton's 3/8 rule, is as follows:

$$\int_a^b f(x)dx = \frac{3h}{8}(f(x_0) + 3f(x_1) + 3f(x_2) + f(x_3)) - \frac{3h^5}{80}f^4(\xi).$$

In all of these formulae the ξ location is inside the interval of the integral $a < \xi < b$. Some numerical texts also list formulae for the $n = 4$ case; however, they are not often used in engineering practice.

6.1.3   Computational example

Let us consider the integral

$$I = \int_a^b f(x)dx = \int_{1/2}^1 x^4 dx$$

for our computational example, since it was instrumental in obtaining the error term for Simpson's rule. We first, however, apply the trapezoid rule.

$$I_t = \frac{1 - 1/2}{2}((1/2)^4 + 1^4) = \frac{17}{64} = 0.2656.$$

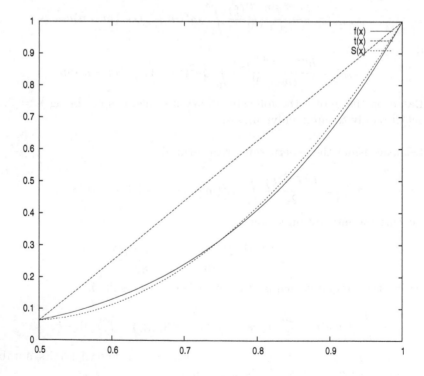

FIGURE 6.1 Numerical integration example

The error term is computed as

$$e_t = \frac{(1/2)^3}{12} f''(\xi) = \frac{1}{8 \cdot 12} 12\xi^2 \le \frac{1}{8} = 0.125.$$

Note that the value of $\xi = 1$ is used to give the maximum. Naturally, the exact result is

$$I = [\frac{x^5}{5}]_{1/2}^1 = \frac{31}{160} = 0.19375.$$

We clearly have an overestimation of the area, since the function is concave from below in the interval of integration. The area between the function $f(x)$ and the trapezoid line $t(x)$ as shown in Figure 6.1 is rather significant.

Now we compute Simpson's rule. The approximate integral is

$$I_S = \frac{1 - 1/2}{2 \cdot 3}((1/2)^4 + 4(3/4)^4 + 1^4) = \frac{1}{12}\frac{169}{64} = 0.19401.$$

This is very clearly much closer to the exact value. The error term for Simpson's rule is

$$e_S = \frac{(1/4)^5}{90} f^{(4)}(\xi) = \frac{24}{90 \cdot 1024} = 0.00026.$$

Simpson's rule is indeed much superior. It is also shown in Figure 6.1 by the very small difference between the function $f(x)$ and the curve noted by $S(x)$ in the figure that represents Simpson's rule.

6.1.4 Open Newton–Cotes formulae

The above forms may all be categorized as closed Newton–Cotes formulae, as the points on the integral boundaries were included ($x_0 = a, x_n = b$), i.e., the interval of integration was closed. Sometimes the function values at the boundaries are not computable, rendering the above formulae useless. For these cases the open Newton–Cotes formulae are used.

To derive the open Newton–Cotes formulae, the following boundary assignment is used:

$$x_0 = a + h$$

and

$$x_n = b - h.$$

Here

$$h = \frac{b - a}{n + 2}$$

is visibly different from the closed formula. The difference is to accommodate the additional inner points required due to the excluded integral boundary points. Following the principle used above, but for a generic n, the open formulae approximation is based on

$$I = \int_{x_0-h}^{x_n+h} \sum_{k=0}^{n} f(x_k)L_k(x)dx.$$

This may also be written as

$$I = \sum_{k=0}^{n} f(x_k) \int_{x_0-h}^{x_n+h} L_k(x)dx.$$

Error terms are computed similarly to the closed formulae, based on the error term of the Lagrange polynomials. The execution of the appropriate integration yields the various open formulae.

The $n = 0$ case is also valid, since we are using $n + 2$ points. This case produces the so-called midpoint formula.

$$\int_a^b f(x)dx = 2hf(x_0) + \frac{h^3}{3}f''(\xi),$$

where the location of the midpoint is

$$x_0 = a + h$$

and the step size is

$$h = \frac{b-a}{2}.$$

The reason for the naming convention is obvious. Similarly, the $n = 1$ open formula is of the form

$$\int_a^b f(x)dx = \frac{3h}{2}(f(x_0) + f(x_1)) + \frac{3h^3}{4}f''(\xi),$$

and the locations of the points are

$$x_0 = a + \frac{b-a}{3}$$

and

$$x_1 = a + 2\frac{b-a}{3} = b - \frac{b-a}{3}.$$

The error formula is related to the same derivative of the function as in the $n = 0$ case, since the latter is considered an "even" case in this regard. Similarly the open formula for $n = 2$ is

$$\int_a^b f(x)dx - \frac{4h}{3}(2f(x_0) - f(x_1) + 2f(x_2)) + \frac{14h^5}{45}f^{(4)}(\xi).$$

6.2 Advanced Newton–Cotes methods

Further generalizations of the Newton–Cotes formulae are possible as shown in this section. These address the issues of very wide integration intervals and possible improvement in the accuracy.

6.2.1 Composite methods

The interval of the integration until now was considered to be narrow enough that a few points were adequate to produce acceptable results. For very wide ranges the composite methods are recommended. The idea dates back to Stirling [12], also in the early 18th century.

Let us now assume that the function to be integrated in a wider range is sampled at $n + 1$ points, with n being even

$$a = x_0 < x_1 < \ldots < x_{n-1} < x_n = b.$$

The points are equidistant, such that

$$x_k = a + kh,$$

with

$$h = (b - a)/n.$$

We apply Simpson's rule to every 3 consecutive points as

$$\int_{x_0}^{x_2} f(x)dx = \frac{h}{3}(f(x_0) + 4f(x_1) + f(x_2)) + e,$$

$$\int_{x_2}^{x_4} f(x)dx = \frac{h}{3}(f(x_2) + 4f(x_3) + f(x_4)) + e,$$

and so on until

$$\int_{x_{n-2}}^{x_n} f(x)dx = \frac{h}{3}(f(x_{n-2}) + 4f(x_{n-1}) + f(x_n)) + e.$$

The complete interval's integral is simply a sum of the intermediate segment's integrals as

$$\int_a^b f(x)dx = \sum_{k=1}^{n/2} \int_{x_{2k-2}}^{x_{2k}} f(x)dx.$$

Substituting Simpson's rule for each intermediate segment's integral results in

$$\int_a^b f(x)dx = \sum_{k=1}^{n/2} (\frac{h}{3}(f(x_{2k-2}) + 4f(x_{2k-1}) + f(x_{2k})) + e).$$

Carrying the summation inside we obtain the composite Simpson's rule of

$$\int_a^b f(x)dx = \frac{h}{3}(f(a) + 2\sum_{k=1}^{n/2-1} f(x_{2k}) + 4\sum_{k=1}^{n/2} f(x_{2k-1}) + f(b)) + \sum_{k=1}^{n/2} e.$$

Some attention to the composite error term is needed. Since

$$e = -\frac{h^5}{90} f^{(4)}(\xi)$$

for each integration segment, the total error is

$$e_C = -\frac{h^5}{90} \sum_{k=1}^{n/2} f^{(4)}(\xi_k).$$

It can be proven by the intermediate value theorem that there is a value such that

$$a < \xi < b,$$

for which

$$nf^{(4)}(\xi) = 2\sum_{k=1}^{n/2} f^{(4)}(\xi_k).$$

With this the error of the composite Simpson's rule is

$$e_{CS} = -\frac{h^5}{180} nf^{(4)}(\xi) = -\frac{(b-a)^5}{180n^4} f^{(4)}(\xi).$$

The same method of derivation produces the composite Newton's 3/8 rule:

$$\int_a^b f(x)dx = \frac{3h}{8}(f(a)+3\sum_{k=1}^{n/3}(f(x_{3k-2})+f(x_{3k-1}))+2\sum_{k=1}^{n/3-1} f(x_{3k})+f(b))+e_N$$

where the error is obtained using an argument similar to that of the composite Simpson's rule:

$$e_N = -\frac{3h^5}{240} nf^{(4)}(\xi) = -\frac{(b-a)^5}{80n^4} f^{(4)}(\xi).$$

Furthermore, the concept carries over to the open formulae as well. The composite midpoint formula may be written as

$$\int_a^b f(x)dx = 2h \sum_{k=0}^{n/2} f(x_{2k}) + \frac{(b-a)^3}{6(n+2)^2} f''(\xi),$$

where

$$h = \frac{b-a}{n+2}.$$

The composite trapezoid formula, quite similarly, is

$$\int_a^b f(x)dx = \frac{h}{2}(f(a) + 2\sum_{k=1}^{n-1} f(x_k) + f(b)) - \frac{(b-a)^3}{12n^2} f''(\xi).$$

This formula is the foundation of the method of the following section.

6.2.2 Romberg's method

Romberg's method [10] is in essence a recursive use of the composite trapezoid formula with a repeated bisectioning of the integration segments. Let us first consider the simple (one segment) trapezoid formula:

$$\int_a^b f(x)dx = \frac{h_1}{2}(f(a) + f(b)) = I_1.$$

We ignore the error terms for now. Here

$$h_1 = b - a.$$

Now apply an $n = 2$ segment (i.e., $n+1 = 3$ point) composite trapezoid rule as

$$\int_a^b f(x)dx = \frac{h_2}{2}(f(a)+2\sum_{k=1}^{n-1} f(x_k)+f(b)) = \frac{h_2}{2}(f(a)+2\sum_{k=1}^{1} f(x_1)+f(b)) = I_2.$$

Note that now

$$h_2 = \frac{b-a}{2} = \frac{h_1}{2},$$

and

$$x_1 = a + h_2 = a + \frac{b-a}{2} = \frac{a+b}{2}.$$

One can see that

$$I_2 = \frac{1}{2}(I_1 + h_1 f(a + h_2)).$$

Further refinement, for example, to $n = 3$ segments, yields

$$I_3 = \frac{1}{2}(I_2 + h_2(f(a + h_3) + f(a + 3h_3))),$$

where now

$$h_3 = \frac{h_2}{2}.$$

This process itself may continue for any k value as

$$I_k = \frac{1}{2}(I_{k-1} + h_{k-1} \sum_{j=1}^{2^{k-2}} (f(a + (2j - 1)h_k))).$$

Finally the approximate integral is

$$\int_a^b f(x)dx = I_k + e_R.$$

The error term was computed originally in [1] and quoted in some advanced numerical tests as

$$e_R = \sum_{j=1}^{\infty} c_j h_k^{2j},$$

where c_j are constants. Clearly, this formula is not really useful in engineering practice. Suffice it to say that several steps of the Romberg refinement are adequate for most practical engineering problems.

We can also observe that the sequence of Romberg integrals is amenable to a Richardson extrapolation. We can assign

$$I_1 = R_1(h),$$

$$I_2 = R_1(h/2),$$

and

$$I_3 = R_1(h/4).$$

The Richardson extrapolation of

$$R_2(h) = R_1(h/2) + \frac{R_1(h/2) - R_1(h)}{3}$$

clearly applies. It is common in engineering practice to execute only a few steps of Romberg's method followed by a Richardson extrapolation to obtain better accuracy results quicker.

6.2.3   Computational example

We demonstrate Romberg's method with the following integral:

$$\int_0^\pi sin(x)dx = 2.$$

This example, like some of the earlier ones, is analytically solvable; however, it was chosen to demonstrate the approximate computation and provide a useful template for the engineer. The simple trapezoid formula is of course

$$I_1 = \frac{\pi}{2}(sin(0) + sin(\pi)) = 0.$$

Following the formula above,

$$I_2 = \frac{1}{2}(I_1 + \pi sin(\pi/2)) = \frac{\pi}{2},$$

and

$$I_3 = \frac{1}{2}(I_2 + \frac{\pi}{2}(sin(\pi/4) + sin(\frac{3\pi}{4}))) = \frac{\pi}{4}(1 + \sqrt{2}).$$

It is clear that the function sampling values may be well organized in a tabular form, as shown in Table 6.1.

TABLE 6.1
Romberg's method

x_j	$k = 1$	$k = 2$	$k = 3$	$k = 4$
0	$sin(0)$			
$\pi/8$				$sin(\frac{\pi}{8})$
$\pi/4$			$sin(\frac{\pi}{4})$	
$3\pi/8$				$sin(\frac{3\pi}{8})$
$\pi/2$		$sin(\frac{\pi}{2})$		
$5\pi/8$				$sin(\frac{5\pi}{8})$
$3\pi/4$			$sin(\frac{3\pi}{4})$	
$7\pi/8$				$sin(\frac{7\pi}{8})$
π	$sin(\pi)$			
I_k	0	1.5708	1.8961	1.9742

The bottom row of Table 6.1 shows the steadily, albeit slowly, increasing accuracy of the approximate integral. We will accelerate the approximation by applying Richardson extrapolation, as shown in Table 6.2.

TABLE 6.2
Romberg's method example

h	$R_1(h)$	$R_2(h)$	$R_3(h)$	$R_4(h)$
π	0			
$\pi/2$	1.5708	2.0944		
$\pi/4$	1.8961	2.0045	1.9986	
$\pi/8$	1.9742	2.0093	1.9999	2.0000

The improvement of the extrapolation is spectacular, actually reaching accurate results within the 4-decimal digit accuracy used. This was achieved without the further need to evaluate the function at other locations.

There are other possible extensions to the Newton–Cotes formulae. One is the class of adaptive quadrature methods. These methods relax the restriction of equidistant intervals. While they are theoretically interesting, the difficulties of the resulting formulae outweigh the advantages gained. The adaptive methods do increase the accuracy, but they do not widen the order of polynomials for which the methods are accurate. The following section describes a class of methods that significantly increase the range of the polynomials for which the method is accurate, a more important feat in engineering applications.

6.3 Gaussian quadrature

The generic closed Newton–Cotes formula using n points is exact for a polynomial of order n or $n + 1$. Gauss investigated the possibility of improving on this by relaxing the restriction of equidistant point selection. The term quadrature comes from area measurement and it was used by Gauss himself. It is also useful to distinguish from the volume integration techniques that are called cubature along the same historical lines.

Specifically, let us seek an approximate integration,

$$\int_a^b f(x)dx = \sum_{i=1}^n c_i f(x_i),$$

that is exact for a polynomial of up to degree $2n - 1$. This condition is described by the equations

$$\sum_{i=1}^{n} c_i = \int_a^b 1 dx = b - a,$$

$$\sum_{i=1}^{n} c_i x_i = \int_a^b x dx = \frac{b^2 - a^2}{2},$$

$$\sum_{i=1}^{n} c_i x_i^2 = \int_a^b x^2 dx = \frac{b^3 - a^3}{3},$$

and so on until

$$\sum_{i=1}^{n} c_i x_i^{2n-1} = \int_a^b x^{2n-1} dx = \frac{b^{2n} - a^{2n}}{2n}.$$

In all we have $2n$ equations and $2n$ unknowns, n of both the c_i and x_i values. The system may be solved for any given a, b; however, a set of precomputed and tabulated solutions exists with $a = -1, b = 1$. The engineer may be able to convert any a, b interval into the precomputed interval by the coordinate transformation

$$x = \frac{a + b}{2} + \frac{b - a}{2} t = g(t).$$

Then the given integral is transformed as

$$\int_a^b f(x) dx = \int_{-1}^{+1} f(g(t)) g'(t) dt,$$

where

$$g'(t) = \frac{b - a}{2}.$$

Hence

$$\int_a^b f(x) dx = \frac{b - a}{2} \int_{-1}^{+1} f(\frac{a + b}{2} + \frac{b - a}{2} t) dt.$$

Finally,

$$\int_a^b f(x) dx = \frac{b - a}{2} \sum_{i=1}^{n} c_i f(\frac{a + b}{2} + \frac{b - a}{2} t_i).$$

Now let us turn our attention to the actual values of c_i, t_i. The c_i are called the Gaussian weights and the t_i are the so-called Gauss points. Without the rather convoluted proof, we state that the Gauss points are the zeroes of the nth order Legendre polynomial, introduced in Section 4.2. The trivial case of $n = 1$ produces

$$Le_1(t) = t = 0,$$

or

$$t_1^1 = 0.$$

For $n = 2$ the t_i are computed from the second Legendre polynomial as

$$Le_2(t) = 3t^2 - 1 = 0,$$

resulting in

$$t_1^2 = -\frac{1}{\sqrt{3}}$$

and

$$t_2^2 = \frac{1}{\sqrt{3}}.$$

Finally for $n = 3$, the third Legendre polynomial zeroes

$$Le_3(t) = \frac{1}{2}(5t^3 - 3t) = 0$$

are at

$$t_1^3, t_3^3 = \pm\sqrt{\frac{3}{5}}$$

and

$$t_2^3 = 0.$$

The superscript introduced here, also used in the computation of the c_i in the following, is to distinguish between the c_i values for different numbers of integration points. The Gauss weights are computed from

$$c_i^n = \int_{-1}^{+1} L_i^n(t)dt,$$

where the L_i^n Lagrange polynomial is using the just computed zeroes of the nth Legendre polynomials as basis points. The Legendre polynomials have their zeroes in the $[-1, 1]$ interval and they are located symmetrically with respect to the origin; these characteristics contribute to their unique role here.

The $n = 1$ case is trivial:

$$c_1^1 = 2.$$

For $n = 2$, the two point integration weights are

$$c_1^2 = \int_{-1}^{+1} L_1^2(t)dt = \int_{-1}^{+1} \frac{t - t_2}{t_1 - t_2}dt = \int_{-1}^{+1} \frac{t - \frac{1}{\sqrt{3}}}{\frac{-1}{\sqrt{3}} - \frac{1}{\sqrt{3}}}dt = 1,$$

and likewise,

$$c_2^2 = 1.$$

Finally for $n = 3$, the weights of the integration are

$$c_1^3 = \int_{-1}^{+1} L_1^3(t)dt = \int_{-1}^{+1} \frac{(t - t_2)(t - t_3)}{(t_1 - t_2)(t_1 - t_3)} dt = \int_{-1}^{+1} \frac{t(t - \sqrt{\frac{3}{5}})}{-\sqrt{\frac{3}{5}}(-\sqrt{\frac{3}{5}} - \sqrt{\frac{3}{5}})} = \frac{5}{9},$$

$$c_2^3 = \int_{-1}^{+1} L_2^3(t)dt = \frac{8}{9},$$

and from symmetry

$$c_3^3 = c_1^3.$$

Higher order formulae may also be produced and are used in engineering practice. The Gauss points and weights up to $n = 6$ with 6-decimal digit accuracy are collected in Table 6.3.

More digits and other quadrature formulae may be found in [7]. The error of the Gaussian quadrature is

$$E_G = \frac{f^{(2n)}(\xi)}{(2n)!} \int_a^b \prod_{i-1}^n (x - x_i)^2 dx,$$

which is not an easily computable formula [8].

6.3.1 Computational example

The technique is first validated with a simple example:

$$\int_{-1}^{+1} x^3 dx = 0.$$

To compute the above integral exactly, the $n = 2$ point formula may be used:

$$\sum_{i=1}^2 c_i^2 f(t_i) = c_1^2(t_1^2)^3 + c_2^2(t_2^2)^3 = 1 \cdot (\frac{-1}{\sqrt{3}})^3 + 1 \cdot (\frac{1}{\sqrt{3}})^3 = 0,$$

which verifies the desirable characteristic of the method. Let us also compute approximately

$$\int_0^1 x^2 e^{-x} dx.$$

TABLE 6.3
Gauss points and weights

n	i	t_i	c_i
1	1	0	2
2	1	−0.577350	1
2	2	0.577350	1
3	1	−0.774597	0.555556
3	2	0	0.888889
3	3	0.774597	0.555556
4	1	−0.861136	0.347855
4	2	−0.339981	0.652146
4	3	0.339981	0.652146
4	4	0.861136	0.347855
5	1	−0.906180	0.236927
5	2	−0.538469	0.478629
5	3	0	0.568889
5	4	0.538469	0.478629
5	5	0.906180	0.236927
6	1	−0.932470	0.171325
6	2	−0.661209	0.360762
6	3	−0.238619	0.467914
6	3	0.238619	0.467914
6	2	0.661209	0.360762
6	1	0.932470	0.171325

Here the integral boundaries need to be converted. Since $a = 0$ and $b = 1$, then

$$g(t) = \frac{1}{2} + \frac{1}{2}t,$$

and

$$dx = \frac{1}{2}dt.$$

The transformed integral becomes

$$\frac{1}{2}\int_{-1}^{+1}\left(\frac{1+t}{2}\right)^2 e^{-\left(\frac{1+t}{2}\right)}\,dt.$$

We will now use the $n = 3$ formula as

$$\frac{1}{2}\left(\frac{5}{9}\left(\frac{1-\sqrt{\frac{3}{5}}}{2}\right)^2 e^{-\left(\frac{1-\sqrt{\frac{3}{5}}}{2}\right)} + \frac{8}{9}\left(\frac{1+0}{2}\right)^2 e^{-\frac{1}{2}} + \frac{5}{9}\left(\frac{1+\sqrt{\frac{3}{5}}}{2}\right)^2 e^{-\left(\frac{1+\sqrt{\frac{3}{5}}}{2}\right)}\right).$$

Evaluating this with 4 decimal digits gives 0.1606. The integral is rather difficult to evaluate analytically, resulting in

$$[e^{-x}(-x^2 - 2x - 2)]_0^1 = 2 - \frac{5}{e} = 0.1606,$$

demonstrating the excellence of the method.

6.4 Integration of functions of multiple variables

We will now view the case of multiple integrals. Mainly we focus on functions of two variables; however, much of the discussion carries over to more than two variables. The goal is to approximately compute the

$$V = \int_a^b \int_{c(x)}^{d(x)} f(x, y)dy \, dx$$

integral. As this is, in essence, the volume between the function and the xy plane, bounded by the planes $x = a$ and $x = b$ and the surfaces $y = c(x)$ and $y = d(x)$, the procedure is sometimes called cubature.

Using Simpson's rule in the inner integral first, we obtain

$$V = \int_a^b \frac{h(x)}{3}(f(x, c(x)) + 4f(x, c(x) + h(x)) + f(x, d(x))dx,$$

where

$$h(x) = \frac{d(x) - c(x)}{2}.$$

Note that $h(x)$ is different from the constant h. Executing Simpson's rule again for the remaining integral yields

$$V = \frac{h}{3}(\frac{h(a)}{3}(f(a, c(a)) + 4f(a, c(a) + h(a)) + f(a, d(a)))$$

$$+ 4\frac{h(a+h)}{3}(f(a+h, c(a+h)) + 4f(a+h, c(a+h)+h(a+h)) + f(a+h, d(a+h)))$$

$$+ \frac{h(b)}{3}(f(b, c(b)) + 4f(b, c(b) + h(b)) + f(b, d(b))) + E,$$

where now the constant h is defined as

$$h = \frac{b - a}{3}.$$

The error term is similarly generalized and the composite Newton–Cotes formulae may be extended for multiple integrals as well.

6.4.1   Gaussian cubature

For functions of two variables, the Gaussian quadrature is extended rather simply into the Gaussian cubature of the form

$$\int_{-1}^{+1} \int_{-1}^{+1} f(x,y) dy dx = \sum_{i=1}^{n} \sum_{j=1}^{n} c_i^n c_j^n f(t_i^n, t_j^n),$$

where the weights and the points are the same as before. For a quick example, the unit height column above the $[-1 \leq x \leq 1, -1 \leq y \leq 1]$ interval

$$V = \int_{-1}^{+1} \int_{-1}^{+1} 1 \, dy \, dx$$

is computed with $n = 2$ as

$$V = \sum_{i=1}^{2} \sum_{j=1}^{2} c_i^n c_j^n = c_1^2(c_1^2 + c_2^2) + c_2^2(c_1^2 + c_2^2) = 1(1+1) + 1(1+1) = 4.$$

Since the function was a constant, the Gaussian locations were not used. This is, of course, the same as the analytic result.

On a final note, the Gauss integral of a function of three variables is

$$\int_{-1}^{+1} \int_{-1}^{+1} \int_{-1}^{+1} f(x,y,z) \, dz \, dy \, dx = \sum_{i=1}^{n} \sum_{j=1}^{n} \sum_{k=1}^{n} c_i^n c_j^n c_k^n f(t_i^n, t_j^n, t_k^n).$$

This formulation is of major importance in the finite element analysis techniques introduced in Chapter 12. For different integration boundaries, the same considerations apply as in the single-variable integral case.

6.5   Chebyshev quadrature

We briefly review the case Chebyshev [3] investigated more than a century ago. His focus was to develop quadrature formulae with constant coefficients for a given number of points n, such that

$$\int_a^b p(x)f(x)dx = c_n \sum_{k=1}^n f(x_k)$$

where $p(x)$ is a weight function continuous in $[a, b]$. Assuming that the formula is at least exact for a constant function, we can establish the constant coefficient as

$$c_n = \frac{1}{n} \int_a^b p(x)dx.$$

Specifically focusing on $p(x) = 1, a = -1, b = 1$, we get

$$\int_{-1}^1 f(x)dx = \frac{2}{n} \sum_{k=1}^n f(x_k).$$

The basis point locations are found as the roots of the following functions:

$$G_1(x) = x,$$

$$G_2(x) = \frac{1}{3}(3x^2 - 1),$$

$$G_3(x) = \frac{1}{2}(2x^3 - x),$$

and so on. The locations of the Chebyshev quadrature points up to $n = 6$ are shown in Table 6.4.

Bernstein in [2] proved that the Chebyshev quadrature formulae exist for $n = 1, \ldots, 7, 9$ and that there is no such formula for $n = 8, 10$ or above. There exist more difficult Chebyshev formulae with different weight functions.

6.6 Numerical integration of periodic functions

For a brief closing section of the discussion on numerical integration, let us consider periodic functions. We seek a special quadrature formula for n points such that

$$\int_0^{2\pi} f(x)dx = \sum_{k=1}^n c_k f(x_k).$$

TABLE 6.4
Chebyshev points and
weights

n	i	t_i	c_n
1	1	0	2
2	1	−0.577350	1
2	2	0.577350	1
3	1	−0.707107	0.666667
3	2	0	0.666667
3	3	0.707107	0.666667
4	1	−0.794654	0.5
4	2	−0.187592	0.5
4	3	0.187592	0.5
4	4	0.794654	0.5
5	1	−0.832497	0.4
5	2	−0.374541	0.4
5	3	0	0.4
5	4	0.374541	0.4
5	5	0.832497	0.4
6	1	−0.866247	0.333334
6	2	−0.422519	0.333334
6	1	−0.266635	0.333334
6	1	0.266635	0.333334
6	2	0.422519	0.333334
6	1	0.866247	0.333334

We consider the following mth order trigonometric function:

$$f(x) = a_0 + \sum_{k=1}^{m}(a_k cos(kx) + b_k sin(kx)).$$

The choice of

$$x_k = (k-1)\frac{2\pi}{n}$$

and

$$c_k = \frac{2\pi}{n}$$

assures an exact integral of up to $m = n - 1$ as

$$\int_0^{2\pi} f(x)dx = \frac{2\pi}{n} \sum_{k=1}^{n} f((k-1)\frac{2\pi}{n}).$$

This may be proven by applying trigonometric identities and evaluating the integral recursively. The formula generalizes for any period T, as

$$\int_0^T f(x)dx = \frac{T}{n} \sum_{k=1}^{n} f((k-1)\frac{T}{n}).$$

The formula is exact for up to $m = n - 1$ order periodic functions of period T:

$$f(x) = a_0 + \sum_{k=1}^{n-1}(a_k cos(\frac{2\pi}{T}kx) + b_k sin(\frac{2\pi}{T}kx)).$$

It is easy to recognize the relationship with a Fourier approximation of functions, a topic of Section 4.4, in the above formulae.

References

[1] Bauer, F. L.; La méthode d'intégration numérique de Romberg, *Colloque sur l'Analyse Numérique*, Librairie Universitaire, Louvain, 1961.

[2] Bernstein, S.; Sur la formule de quadrature approchée de Tschebycheff, *Comp. Rend. Acad. Sci.*, Vol. 203, pp. 1305–1306, 1936.

[3] Chebyshev, P. L.; Sur les quadratures, *J. Math. Pures Appl*, Vol. 19, pp. 19–34, 1874.

[4] Cotes, R.; De methodo differentiali Newtoniana, *Harmonia Mensurarum*, Cambridge, 1722.

[5] Gauss, F,; Methodus nova integralium valores per approximationem inveniendi, *Koniglichen Geselschaft des Wissenshaften*, Gottingen, Vol. 4, pp. 163–196, 1866.

[6] Issacson, E. and Keller, H. B.; *Analysis of Numerical Methods*, Wiley, New York, 1966.

[7] Konrod, A. S.; *Nodes and Weights for Quadrature Formulae. Sixteen Place Tables*, Consultants Bureau, New York, 1965.

[8] Krylov, V. I.; Approximate calculation of integrals, *Izdat. Fiz.-Mat. Lit.*, Moscow, 1959.

[9] Newton, I.; *Philosophiae Naturalis Principia Mathematica*, London, 1687.

[10] Romberg, W.; Vereinfahcte Numerische Integration, *Kgl. Norske Vid. Selsk. Forsk.*, Vol. 28, pp. 30–36, 1955.

[11] Simpson, T.; *Mathematical Dissertations on a Variety of Physical and Analytical Subjects*, London, 1743.

[12] Stirling, J.; Methodus differentialis Newtoniana illustrata, *Phil. Trans. Royal Society*, Vol. 30, pp. 1050–1070, 1718.

Part II

Approximate solutions

7

Nonlinear equations in one variable

The topic of Part II, solution approximations, is just as familiar to engineers as the topic of Part I, interpolation. It is also an area of long history; the interest in solving algebraic equations dates back to the ancient geometers. Closed-form solutions for algebraic equations of higher order were actively sought throughout the Middle Ages, culminating in Cardano's formulae for cubic and quartic equations.

The earliest techniques for generic, non-algebraic equations are from Raphson [6], still in the 17th century. The understanding of the limits of analytical solution of higher order algebraic equations came with Ruffini in 1799 [7] and with the formal proof of Abel in 1826 [1]. Hence the interest focused on the approximate solution of higher order algebraic equations, resulting in a wealth of theoretical results related to existence and intervals of solutions [2], [8].

This chapter, true to its focus on engineering computations, describes the concepts and methods most practical in this regard. More details may be found in numerical analysis texts such as [5].

7.1   General equations

We first focus on the case of

$$f(x) = 0,$$

where $f(x)$ is general, but neither linear, nor necessarily a polynomial function.

7.1.1   The method of bisection

The simplest method of finding approximate roots of a general equation is the method of bisection. Assume that we have narrowed down an interval $[a, b]$ such that it contains only one real root. In this case,

$$sign(f(a)) \neq sign(f(b))$$

applies. The method starts as

$$p_1 = a + \frac{b-a}{2}.$$

If

$$sign(f(p_1)) = sign(f(b)),$$

then choose

$$b_1 = p_1, a_1 = a;$$

otherwise select

$$a_1 = p_1, b_1 = b.$$

The method continues by generating the following sequence:

$$p_i = a_{i-1} + \frac{b_{i-1} - a_{i-1}}{2}$$

and checking the

$$sign(f(p_i)) = sign(f(b_{i-1}))$$

relationship to set either

$$b_i = p_i, a_i = a_{i-1},$$

or

$$a_i = p_i, b_i = b_{i-1}.$$

The process stops when the

$$\frac{|p_i - p_{i-1}|}{|p_i|}$$

quantity that is the relative error of the approximation is sufficiently small.
Then

$$f(p_i) = 0 + E,$$

where E is the error of the method after i number of steps have been executed.
For these types of methods, the error can always be made smaller by executing more steps. It is also possible that the method finds an exact solution in a finite number of steps, a welcome event indeed. View the first steps of the bisection process:

$$b_1 - a_1 = \frac{b-a}{2},$$

$$b_2 - a_2 = \frac{b_1 - a_1}{2} = \frac{b - a}{4}.$$

Hence

$$b_i - a_i = \frac{b - a}{2^i}.$$

Assuming that the exact root is r,

$$|p_i - r| \leq \frac{b_i - a_i}{2} = \frac{b - a}{2^i}.$$

This indicates that having a tight starting interval is imperative for the success of this method. Another interesting issue is the rate of convergence to the approximate solution, written as

$$p_i = r + O(\frac{1}{2^i}).$$

This method, while conceptually very simple, and frequently used in quick approximate computations by engineers, is not very efficient.

7.1.2 The secant method

The starting environment in this case is again

$$sign(f(a)) \neq sign(f(b)).$$

The geometric concept of this technique's approximation is to connect the two points at the end of the interval with a secant line and choose the next point of iteration to be the intersection of the secant line with the x-axis. The process is shown in Figure 7.1, where $s(x)$ denotes the secant lines and $f(x)$ is the functions whose zero is sought.

The equation of the secant line for $i = 0, 1, 2, \ldots$ is

$$y - f(x_a^i) = \frac{f(x_b^i) - f(x_a^i)}{x_b^i - x_a^i}(x - x_a^i),$$

where

$$x_a^0 = a, x_b^0 = b.$$

The sequence of approximate roots is

$$q_i = x_a^i - f(x_a^i)\frac{x_b^i - x_a^i}{f(x_b^i) - f(x_a^i)}.$$

As in the bisection method, we judiciously replace one of the boundary points with the newly found point. If

$$sign(f(q_i)) = sign(f(x_b^i)),$$

then

$$x_b^{i+1} = q_i, x_a^{i+1} = x_a^i;$$

otherwise

$$x_a^{i+1} = q_i, x_b^{i+1} = x_b^i.$$

The sequence of q_i values approximates the root

$$q_i \approx r,$$

when

$$\frac{|q_i - q_{i-1}|}{|q_i|}$$

is small enough. Since the function values are readily available at each iteration step, the alternative stopping criterion of

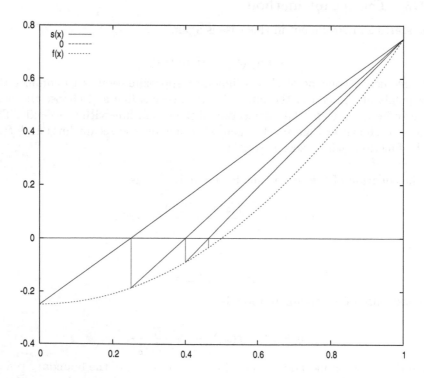

FIGURE 7.1 Concept of the secant method

$$|f(q_i)| < \epsilon$$

may also be used, where ϵ is a small number established by the engineer's desire for accuracy. This method, like the bisection method, converges unconditionally as long as the initial condition of the interval containing one root is satisfied.

7.1.3 Fixed point iteration

There is yet another extension of these types of methods, the method of fixed point iteration. This method relies on the possible decoupling of a term from the equation as

$$f(x) = g(x) - x.$$

Based on that decoupling, the original

$$f(x) = 0$$

problem is replaced by

$$g(x) = x.$$

Assuming a reasonable estimate x_0 for the root, a simple iterative process is executed by

$$x_1 = g(x_0),$$
$$x_2 = g(x_1),$$

and

$$x_k = g(x_{k-1}).$$

We assume that the exact root is r and the approximate root is in a small δ neighborhood of the exact root

$$|x_0 - r| < \delta.$$

We introduce

$$|g'(x)| \le q.$$

Then

$$|x_1 - r| = |g(x_0) - g(r)| = |g'(\xi_1)||x_0 - r| \le q\delta.$$

The next step produces

$$|x_2 - r| = |g(x_1) - g(r)| = |g'(\xi_2)||x_1 - r| \le q^2\delta,$$

and so on until
$$|x_i - r| \le q^i \delta.$$
This implies that

$$lim_{i \to \infty} |x_i - r| = 0,$$
if
$$|g'(x)| \le q < 1.$$

This is the convergence criterion for this sequence. It may preclude this method's use in some cases, even if the required decoupling is attainable.

7.1.4   Computational example

To demonstrate the above three basic methods, we consider the problem of finding an approximate solution to the

$$f(x) = x^2 - \frac{1}{2} = 0$$

equation in the interval of $[0,1]$. The problem, of course, may be solved with well-known analytic tools (the exact root is $\sqrt{2}/2$), but its simplicity is intended to enlighten the procedures. We will execute two steps of each of the three methods introduced in this section.

The bisection method starts from

$$a_0 = 0, b_0 = 1, f(a_0) = -\frac{1}{2}, f(b_0) = \frac{1}{2},$$

and proceeds with first finding the bisection value of $p_1 = \frac{1}{2}$ at which point the function is valued at $-\frac{1}{4}$. Hence this is negative, the selection is

$$a_1 = \frac{1}{2}, b_1 = 1.$$

The second bisection step yields $p_2 = \frac{3}{4}$ and the function value of $\frac{1}{16}$. We will accept that as an approximate root. The difference between the analytical and approximate solution is 0.04.

The secant method starting also from the same interval,

$$x_a^0 = 0, x_b^0 = 1,$$

finds the first secant intersection at $q_0 = \frac{1}{2}$, which is still the same as in the bisection method. The second secant step produces $q_1 = \frac{2}{3}$, at which point the function is $f(q_1) = -\frac{1}{18}$. The difference between the analytical and approximate solution is also about 0.04, similar to that of the bisection method.

In order to use the fixed point method, a small amount of heuristics is required. Since there is no clearly visible way to decouple an x term from the example function, the trick of adding or subtracting x on both sides of $f(x) = 0$ is commonly applied. In this example, the subtraction produces

$$x = x - x^2 + \frac{1}{2} = g(x)$$

and

$$g'(x) = 1 - 2x,$$

which is less than one inside the interval of interest, as desired, albeit not on the boundary. Therefore we start from the interior point

$$x_0 = \frac{1}{2}$$

and produce

$$x_1 = \frac{1}{2} - \frac{1}{4} + \frac{1}{2} = \frac{3}{4}.$$

At this point the function value is already $\frac{1}{16}$. Executing the second step of the fixed point iteration results in an approximate root of

$$x_2 = \frac{3}{4} - \frac{9}{16} + \frac{1}{2} = \frac{11}{16}$$

and a function value of

$$f(x_2) = -\frac{7}{256}.$$

The difference between the analytical and approximate solution is about 0.02, which is twice as good as the secant or the bisection method. This is a good result, even if we consider the fact that we have started from an internal point of the interval and some heuristics were required.

7.2 Newton's method

Newton's is a method with a faster rate of convergence. In order to be applied, however, this method requires the function to be twice differentiable.

To start, a location satisfying the following condition is needed:

$$sign(f(x_0)) = sign(f'(x_0)),$$

where

$$a \leq x_0, r \leq b,$$

and r is the exact root. The geometric concept is again rather simple; we use the zero of the tangent line from that starting point to provide the next iteration. In Figure 7.2 the tangent lines are denoted by $t(x)$ along with the associated vertical line segment used to find the next point on the function $f(x)$. The process algebraically starts as

$$x_1 = x_0 - \frac{f(x_0)}{f'(x_0)}$$

and continues by repeating the step

FIGURE 7.2 Concept of Newton's method

$$x_i = x_{i-1} - \frac{f(x_{i-1})}{f'(x_{i-1})}.$$

It may be proven that the sequence of

$$x_0, x_1, x_2, \ldots, x_i$$

is monotonic and converges to the exact root

$$lim_{i \to \infty} x_i = r.$$

Substituting the formula

$$lim_{i \to \infty} (x_{i-1} - \frac{f(x_{i-1})}{f'(x_{i-1})}) = r - \frac{f(r)}{f'(r)}.$$

The above limit is r if $f(r) = 0$. The rate of convergence of Newton's method, as promised at the beginning of this chapter, is very good. Consider the Taylor polynomial of the function in the neighborhood of the last iterate, but ignore the truncation error term

$$f(x) = f(x_{i-1}) + (x - x_{i-1})f'(x_{i-1}) + \frac{1}{2}(x - x_{i-1})^2 f''(x_{i-1}).$$

Substituting $x = r$, for which $f(r) = 0$ yields

$$0 = f(x_{i-1}) + (r - x_{i-1})f'(x_{i-1}) + \frac{1}{2}(r - x_{i-1})^2 f''(x_{i-1}).$$

By reordering we obtain

$$r - (x_{i-1} - \frac{f(x_{i-1})}{f'(x_{i-1})}) = -\frac{1}{2}(r - x_{i-1})^2 \frac{f''(x_{i-1})}{f'(x_{i-1})}.$$

Substituting x_i yields a formula for the rate of convergence of the method,

$$\frac{r - x_i}{(r - x_{i-1})^2} = -\frac{1}{2}\frac{f''(x_{i-1})}{f'(x_{i-1})},$$

where the right-hand side is a constant, say, c. Hence

$$|r - x_i| = c(r - x_{i-1})^2.$$

The distance of the ith iterate from the root is proportional to the square of the distance of the previous iterate. Hence the rate of convergence of Newton's method is quadratic, a significant fact indeed. On the other hand, the method is clearly ineffective in a neighborhood of points where the derivative is nearly horizontal.

Newton's method applied to the simple example

$$f(x) = x^2 - \frac{1}{2} = 0$$

proceeds as follows. Starting from

$$x_0 = 1$$

the first iterations produce

$$x_1 = 1 - \frac{1/2}{2} = \frac{3}{4}.$$

The second step is

$$x_2 = 3/4 - \frac{1/16}{3/2} = \frac{17}{24}.$$

The function value at this point is 1/576, which prompts us to accept this as an approximate root. The difference between the analytical and approximate solution is 0.001. The superiority of the method over the classical methods of the last section is well demonstrated by having an excellent result after only two steps.

The method has a simplified version, called the Newton–Raphson [6] method, which removes the need for repeated evaluation of the derivative. The form is simply

$$x_i = x_{i-1} - \frac{f(x_{i-1})}{f'(x_0)},$$

meaning the derivative at the starting point is used repeatedly. The price of the easement is a reduction in convergence rate.

A modification in the other direction, using the second derivative, has been proposed by Householder [3] in the form of

$$x_i = x_{i-1} - \frac{f(x_{i-1})}{f'(x_{i-1})}\left(1 + \frac{f(x_{i-1})f''(x_{i-1})}{2(f'(x_{i-1}))^2}\right).$$

Finally an extension of Newton's method for the case of complex roots was developed by Laguerre [4], but it is not detailed here.

Despite the seemingly stricter starting conditions, Newton's method is probably the most widely and successfully used by engineers. It can also be extended to the solution of systems of nonlinear equations, to be discussed in the next chapter.

7.3 Solution of algebraic equations

The problem at hand is to find approximate solutions of

$$f(x) = \sum_{i=0}^{n} a_i x^i = 0,$$

where the a_i coefficients of the polynomial of order n are real. Some of the methods discussed here will also apply for the case of complex coefficients. From the teaching of calculus it is well known that between consecutive zeroes of this equation the derivative function $f'(x)$ also has at least one root. It follows that if the function has m roots in an interval $[a, b]$, then the derivative has at least $m - 1$ zeroes in the same interval.

7.3.1 Sturm sequence

The Euclidean algorithm for two polynomials of $f(x)$ and $g(x)$ is started as

$$f(x) = g(x)q_0(x) + r_1(x),$$
$$g(x) = r_1(x)q_1(x) + r_2(x).$$

Then

$$r_1(x) = r_2(x)q_2(x) + r_3(x),$$

and so on, until

$$r_{m-2}(x) = r_{m-1}(x)q_{m-1}(x) + r_m,$$
$$r_{m-1}(x) = r_m(x)q_m(x).$$

The process stops when r_m becomes a constant. If $r_m = 0$, then $r_{m-1}(x)$ is the greatest common divisor of the two polynomials. If $r_m \neq 0$, then the two polynomials do not have a common divisor.

We apply this algorithm for the function and its derivative as follows:

$$f(x) = f'(x)q_0(x) - r_1(x),$$
$$f'(x) = r_1(x)q_1(x) - r_2(x).$$

Then

$$r_1(x) = r_2(x)q_2(x) + r_3(x),$$

and so on, until

$$r_{m-2}(x) = r_{m-1}(x)q_{m-1}(x) - r_m,$$

$$r_{m-1}(x) = r_m(x)q_m(x).$$

The negative sign in the sequence indicates that we take the residuals with the opposite sign. The sequence

$$f, f', r_1, r_2, \ldots, r_m$$

is a so-called Sturm sequence. Assuming that $f(x)$ has simple roots only, the last residual will be different from zero. Sturm's theorem [8] states that the number of zeroes of $f(x)$ in the interval $[a, b]$ is equal to the number

$$V(a) - V(b),$$

where $V(a)$ is the number of sign changes in the sequence above for $x = a$ and $V(b)$ is the same for $x = b$. This is correct in the case of multiple roots as well, as long as they are counted as one, i.e., the number of different roots is found by the above theorem.

Finally, the theorem may be extended into the case of complex functions and complex domains. These could be of interest in investigating the stability of mechanical systems. For a stable system all roots must have a negative real part. Without immersing ourselves in more details, let the following theorem be stated: a complex function with real coefficients

$$f(z) = \sum_{i=0}^{n} a_i z^i, a_n > 0$$

has only roots in the left-hand complex half plane if all the coefficients are positive. There are many other theorems concentrating on conditions for the roots being in a complex circle that are not detailed here. In the remainder of this chapter we will focus on the real case unless otherwise stated.

7.3.2   Horner's scheme of evaluating polynomials

Before we can use any of the earlier methods to find an approximation of a root in one of the intervals established by Sturm's theorem, we must focus on minimizing the drudgery of evaluating a polynomial.

A polynomial with a nonzero leading coefficient of a_n,

$$f(x) = a_n x^n + a_{n-1} x^{n-1} + \ldots + a_1 x + a_0,$$

may be reordered as

$$f(x) = (((\ldots(a_n x + a_{n-1})x + a_{n-2})x + \ldots)x + a_1)x + a_0.$$

In order to evaluate the function at x_0, we recursively proceed from the inside out:

$$b_n = a_n,$$

$$b_{n-1} = b_n x_0 + a_{n-1},$$

$$b_{n-i} = b_{n-i+1} x_0 + a_{n-i},$$

and so on until

$$b_0 = b_1 x_0 + a_0 = f(x_0).$$

The process is conveniently executed in tabulated form, as shown in the next section in connection with a computational example.

There is an extension of Horner's scheme to find the complex conjugate pairs of algebraic equations. The method is called Bairstow's technique and it is not detailed here but can be found, for example, in [4].

7.3.3 Computational example

To demonstrate the use of the Sturm sequence in practice, we consider the fifth order polynomial

$$f(x) = x^5 - x^4 - 3x^3 + 2x + 5.$$

The derivative is

$$f'(x) = 5x^4 - 4x^3 - 9x^2 + 2.$$

In each step a multiplier may be found to avoid dealing with uncomfortable fractions; this leads to the emergence of an interesting pattern that could be further exploited in the case of very high order polynomials. For example, the next member of the Sturm sequence is

$$r_1(x) = \frac{1}{5^2}(34x^3 + 9x^2 - 40x - 127).$$

Continuing the sequence using the same concept in each step results in

$$r_2(x) = \frac{5^2}{34^2}(79x^2 - 574x + 827),$$

$$r_3(x) = \frac{34^2}{5^2 79^2}(-7906 + 15,156).$$

Finally, in the last step we do not need to compute beyond the fact that the member is a negative constant:

$$r_4 = -c.$$

The execution of the evaluation of the Sturm sequence is facilitated by Table 7.1, which allows several observations.

TABLE 7.1

Sturm sequence example

x	$f(x)$	$f'(x)$	$r_2(x)$	$r_3(x)$	$r_4(x)$	$r_5(x)$	$V(x)$
$-\infty$	$-$	$+$	$-$	$+$	$+$	$-$	4
-2	$-$	$+$	$-$	$+$	$+$	$-$	4
-1	$+$	$+$	$-$	$+$	$+$	$-$	3
0	$+$	$+$	$-$	$+$	$+$	$-$	3
1	$+$	$-$	$-$	$+$	$+$	$-$	3
2	$+$	$+$	$+$	$-$	$-$	$-$	1
$+\infty$	$+$	$+$	$+$	$+$	$-$	$-$	1

Since

$$V(1) - V(2) = 2,$$

there are 2 roots between $x = 1$ and $x = 2$. On the other hand,

$$V(-2) - V(-1) = 1,$$

implies a single root in this subinterval. Furthermore, since

$$V(-\infty) - V(+\infty) = 3,$$

the polynomial has only 3 real roots.

We evaluate the function at $x_0 = -1$ using Horner's scheme in Table 7.2. The second and last rows are the original and the modified coefficients, respectively, and the middle row contains the intermediate $b_{i+1} \cdot x_0$ terms for $i = 4, 3, 2, 1, 0$ except that the very first term is zero by definition.

The process results in $f(-1) = 4$, which may be verified by directly substituting into the original polynomial.

TABLE 7.2
Horner's scheme example

i		5	4	3	2	1	0
a_i		1	-1	-3	0	2	5
$x_0 = -1$	0	-1	2	1	-1	-1
b_i		1	-2	-1	1	1	4

7.4 Aitken's acceleration

The concept of improving the approximation has already emerged in the Richardson extrapolation as well as in Romberg's method. Here the technique of accelerating the convergence of a given method is considered.

Let us consider a sequence of iterative values

$$p_0, p_1, p_2, \ldots, p_i$$

that converges to the exact root r. The rate of convergence is measured by the ratio of

$$\frac{r - p_{i+1}}{r - p_i}.$$

Let us assume that we are far enough into the iterative sequence that the condition

$$\frac{r - p_{i+1}}{r - p_i} \approx \frac{r - p_{i+2}}{r - p_{i+1}}.$$

also holds. From this it follows that

$$(r - p_{i+1})^2 \approx (r - p_{i+2})(r - p_i).$$

Executing the posted operations and solving results in

$$r \approx \frac{p_{i+2}p_i - p_{i+1}^2}{p_{i+2} - 2p_{i+1} + p_i}.$$

Some inventive algebra by adding and subtracting terms yields

$$r \approx p_i - \frac{(p_{i+1} - p_i)^2}{p_{i+2} - 2p_{i+1} + p_i}.$$

This proposes Aitken's acceleration, which states that the sequence

$$\bar{p}_i = p_i - \frac{(p_{i+1} - p_i)^2}{p_{i+2} - 2p_{i+1} + p_i}$$

converges faster than the original sequence.

To facilitate an easy computation of the new sequence, we can again apply forward differences to the members of the sequence, as follows:

$$\Delta p_i = p_{i+1} - p_i.$$

This is conceptually similar to, but slightly different from, the forward divided differences introduced earlier. The second order forward difference is

$$\Delta^2 p_i = \Delta(\Delta p_{i+1} - \Delta p_i) = \Delta^2 p_{i+1} - \Delta^2 p_i$$

$$= (p_{i+2} - p_{i+1}) - (p_{i+1} - p_i) = p_{i+2} - 2p_{i+1} + p_i.$$

With these simplifications, the accelerated series can be written as

$$\bar{p}_i = p_i - \frac{(\Delta p_i)^2}{\Delta^2 p_i},$$

which facilitates an orderly computation.

7.4.1   Computational example

Let us consider the sequence

$$p_i = \frac{1}{i},$$

obtained by an iteration scheme shown in the second column of Table 7.3. The sequence is converging to $r = 0$ in a rather slow manner. The third and fourth columns are demonstrating the computation of the forward differences.

TABLE 7.3
Aitken's acceleration
example

i	p_i	Δp_i	$\Delta^2 p_i$	$\bar{p}_i$
1	1/1	$-1/2$	1/3	1/4
2	1/2	$-1/6$	1/12	1/6
3	1/3	$-1/12$	1/30	1/8
4	1/4	$-1/20$		
5	1/5			
6	1/6			
7	1/7			
8	1/8			

The accelerated sequence of $\bar{p}_i$, shown in the last column, clearly demonstrates the advantage of Aitken's method. The first Aitken's term computed is the same as the fourth term of the original sequence, the second is the same as the sixth, and the third term of the accelerated sequence is the same as the eighth term of the original sequence. This conceptually simple and easily computable method is very powerful.

References

[1] Abel, N.; Beweis der Unmöglichkeit algebraische Gleichungen von höheren Graden als dem vierten allgemein aufzulösen, *Grelles Journal*, Vol. 1, p. 65, 1826.

[2] Fourier, J.; *Analyse des equations determines*, Livre I, Paris, 1831.

[3] Householder, A. S.; *The Numerical Treatment of a Single Nonlinear Equation*, McGraw-Hill, New York, 1970.

[4] Press, W. H. et al.; *Numerical Recipes in FORTRAN: The Art of Scientific Computing*, Cambridge University Press, 1989.

[5] Ralston, A.; *A First Course in Numerical Analysis*, McGraw-Hill, New York, 1965.

[6] Raphson, J,; *Analysis equationum universalis*, London, 1690.

[7] Ruffini, P.; *Teoria generale delle equazioni in cui si dimonstra impossibile la soluzione algebraica delle equazioni generali di grado superiore al quarto*, Bologna, 1799.

[8] Sturm, J.; Memoire sur la resolution des equations numeriques, *Bulletin Ferussac*, Paris, 1829.

8

Systems of nonlinear equations

We now turn our attention to a system of nonlinear equations. The topic's history is again related to Newton; several aspects of it were researched by him, although not formally published. In fact most solutions were based on his work until various flavors of a more efficient approach were proposed by the quartet of Broyden [1], Fletcher [2], Goldfarb [3] and Shanno [5] in 1970.

Our general problem is posed in terms of a vector valued function,

$$\underline{f}(\underline{x}) = \underline{0},$$

where the underlining indicates a vector of R^n,

$$\underline{x} = \begin{bmatrix} x_1 \\ x_2 \\ \dots \\ x_n \end{bmatrix},$$

and

$$\underline{f}(\underline{x}) = \begin{bmatrix} f_1(x_1, x_2, \dots, x_n) \\ f_2(x_1, x_2, \dots, x_n) \\ \dots \\ f_n(x_1, x_2, \dots, x_n) \end{bmatrix}.$$

8.1 The generalized fixed point method

The first approach we take to solve the problem of systems of nonlinear equations is by a generalization of the fixed point method introduced in the last chapter. The definition of a fixed point in this sense is that the vector $\underline{p}$ is a fixed point of the vector valued function $\underline{g} \in R^n$ if

$$\underline{g}(\underline{p}) = \underline{p}.$$

The conditions of the existence of a fixed point are stated next. Define an n-dimensional domain D as

$$a_k \le x_k \le b_k; k = 1, 2, \dots, n.$$

If the function $\underline{g}$ meets the condition of

$$g(\underline{x}) \in D; \underline{x} \in D,$$

then the function has a fixed point in D. Let us assume that the nonlinear function is obtained by the partitioning of our nonlinear system as

$$\underline{f}(\underline{x}) = \underline{g}(\underline{x}) - \underline{x} = \begin{bmatrix} g_1(x_1, x_2, \dots, x_n) \\ g_2(x_1, x_2, \dots, x_n) \\ \dots \\ g_n(x_1, x_2, \dots, x_n) \end{bmatrix} - \begin{bmatrix} x_1 \\ x_2 \\ \dots \\ x_n \end{bmatrix}.$$

Furthermore, we also assume that the partial derivatives are all continuous and the matrix of derivatives

$$G(\underline{x}) = \begin{bmatrix} \frac{\partial g_1(\underline{x})}{\partial x_1} & \frac{\partial g_1(\underline{x})}{\partial x_2} & \cdots & \frac{\partial g_1(\underline{x})}{\partial x_n} \\ \frac{\partial g_2(\underline{x})}{\partial x_1} & \frac{\partial g_2(\underline{x})}{\partial x_2} & \cdots & \frac{\partial g_2(\underline{x})}{\partial x_n} \\ \dots & \dots & \dots & \dots \\ \frac{\partial g_n(\underline{x})}{\partial x_1} & \frac{\partial g_n(\underline{x})}{\partial x_2} & \cdots & \frac{\partial g_n(\underline{x})}{\partial x_n} \end{bmatrix}$$

has $det(G) \neq 0$ if $\underline{x} \in D$. Finally, assume that for all $k = 1, 2, \dots, n$ and $j = 1, 2, \dots, n$ the partial derivatives are bounded as

$$|\frac{\partial g_k(\underline{x})}{\partial x_j}| \leq K < 1.$$

Under these conditions the generalized fixed point iteration scheme of

$$\underline{x}_i = \underline{g}(\underline{x}_{i-1}); i = 1, 2, \dots$$

is guaranteed to converge to the fixed point. The rate of convergence is

$$||\underline{x}_i - \underline{p}|| \leq \frac{K^i}{1-K} ||\underline{x}_i - \underline{x}_0||,$$

where the vector norms are measured by the maximum absolute value term. The improvement obtained in the ith step of the iteration is

$$||\underline{x}_i - \underline{p}|| \leq \frac{K}{1-K} ||\underline{x}_i - \underline{x}_{i-1}||.$$

Hence the stopping criterion of the iteration is when the right-hand quantity dips below an acceptable threshold.

$$||\underline{x}_i - \underline{x}_{i-1}|| < \epsilon.$$

An improvement to the generalized fixed point method is via the Seidel concept also used in the Gauss–Seidel method discussed in the next chapter. The concept utilizes the already computed components of the next iterate during the iteration as follows:

$$x_{1,i+1} = g_1(x_{1,i}, x_{2,i}, \ldots, x_{n,i}),$$

$$x_{2,i+1} = g_2(x_{1,i+1}, x_{2,i}, \ldots, x_{n,i}),$$

and so on, until

$$x_{n,i+1} = g_n(x_{1,i+1}, x_{2,i+1}, \ldots, x_{n-1,i+1}, x_{n,i}).$$

The generalized fixed point method is useful for problems in which the component functions are polynomials in multiple variables. Other classes of problems amenable to the generalized fixed point method are those with component functions containing mixed polynomial and trigonometric expressions. In these cases there are some guarantees for continuity and boundedness, which are conditions of the method's convergence.

8.2 The method of steepest descent

While the methods of the latter part of this chapter are the most widely used in engineering practice, the method of steepest descent is important to discuss as it points to a direction that will be used many times in the following chapters.

The problem of the solution of a system of nonlinear equations may be recast in the form of a minimization problem. The solution of the problem is equivalent to finding the minimum of the function

$$g(x) = \sum_{i=1}^{n} (f_i(x))^2.$$

If the global minimum of this function is positive, then the original nonlinear system does not have a solution. If the global minimum is zero, then the point of minimum is the solution of the nonlinear system.

Consider the starting solution of x_0 and the gradient at this point defined by

$$\nabla g(x_0) = \begin{bmatrix} \frac{\partial g(x)}{\partial x_1} \\ \frac{\partial g(x)}{\partial x_2} \\ \ldots \\ \frac{\partial g(x)}{\partial x_n} \end{bmatrix}_{x=x_0}.$$

The function's descent is steepest (hence the name) in the direction opposite to the gradient. We normalize the gradient vector to unit length

$$n(\underline{x}_0) = \frac{\nabla g(\underline{x}_0)}{||\nabla g(\underline{x}_0)||}.$$

Hence an improved solution may be found along the direction

$$\underline{x}(t) = \underline{x}_0 - t \cdot n(\underline{x}_0),$$

where the t parameter is yet unknown. We next find the minimum of the $g(\underline{x})$ function in that direction:

$$g(\underline{x}_0 - t \cdot n(\underline{x}_0)) = min,$$

which is at the location where the derivative with respect to the parameter vanishes, i.e.,

$$\frac{d}{dt} g(\underline{x}_0 - t \cdot n(\underline{x}_0)) = 0.$$

We denote the smallest positive solution as t_0 and find the next iterative solution at

$$\underline{x}_1 = \underline{x}_0 - t_0 \cdot n(\underline{x}_0).$$

The process may be continued as

$$\underline{x}_i = \underline{x}_{i-1} - t_i \cdot n(\underline{x}_i),$$

resulting in an ever decreasing series of function values

$$g(\underline{x}_i) < g(\underline{x}_{i-1}) < \ldots < g(\underline{x}_1) < g(\underline{x}_0).$$

In general,

$$lim_{i \to \infty} g(\underline{x}_i) = 0.$$

If this limit exists, but it is not zero, the method converges to a local minimum of the g function and the original problem has no solution.

There is the issue of the efficient calculation of the t parameter value. The theoretical minimum of the t_i value is rather expensive to compute. Instead we interpolate the

$$h(t) = g(\underline{x}_0 - t \cdot n(\underline{x}_0))$$

function at three suitable distinct locations, say, $t = a, b, c$. We evaluate the function to be minimized at these locations as

$$h(a), h(b), h(c).$$

Then we construct a quadratic polynomial through these points with Newton's forward divided difference formula to approximate $h(t)$,

$$h(t) = h(a) + f[a,b](t-a) + f[a,b,c](t-a)(t-b),$$

which we solve for the pseudo-optimal t value. For more details on solving for a pseudo-optimal t value the reader may consult [6].

8.2.1 Computational example

We use the nonhomogeneous nonlinear system of

$$x_1^2 + x_2 - 3 = 0,$$

and

$$x_1 + x_2^2 - 5 = 0,$$

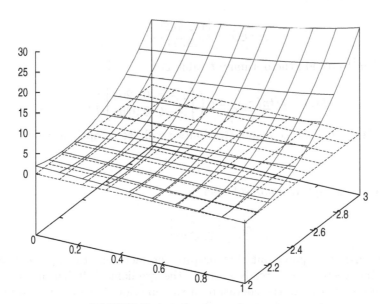

FIGURE 8.1 Steepest descent example

with the exact solution of

$$\underline{x} = \begin{bmatrix} x_1 \\ x_2 \end{bmatrix} = \begin{bmatrix} 1 \\ 2 \end{bmatrix}.$$

For this example,

$$g(x) = (x_1^2 + x_2 - 3)^2 + (x_1 + x_2^2 - 5)^2.$$

The function is shown in Figure 8.1 along with the plane of the solution, denoted by g(1,2). The solution is the $(1, 2)$ point. The gradient is

$$\nabla g(\underline{x}) = \begin{bmatrix} 2(x_1^2 + x_2 - 3)2x_1 + 2(x_1 + x_2^2 - 5) \\ 2(x_1^2 + x_2 - 3) + 2(x_1 + x_2^2 - 5)2x_2 \end{bmatrix}.$$

We will now use the starting location of

$$\underline{x}_0 = \begin{bmatrix} 1 \\ 3 \end{bmatrix},$$

at which the gradient is

$$\nabla g(\underline{x}_0) = \begin{bmatrix} 14 \\ 62 \end{bmatrix}.$$

The normalized vector is

$$\underline{n}(\underline{x}_0) = \begin{bmatrix} 0.2203 \\ 0.9754 \end{bmatrix}.$$

The function to obtain the optimum t value is

$$h(t) = g(t\underline{n}(\underline{x}_0)) = g\left(\begin{bmatrix} 1 - t \cdot 0.2203 \\ 3 - t \cdot 0.9754 \end{bmatrix}\right).$$

For the sake of this simple demonstration we will use as optimal value

$$t = 1,$$

and find the next approximate solution in

$$\underline{x}_1 = \begin{bmatrix} 0.7797 \\ 2.0246 \end{bmatrix}.$$

It is important to point out that despite the fact that the x_1 coordinate of the initial vector was already correct, the gradient shift has moved the next iterate away, due to the curvature of the function. Several more steps of "shooting over" the desired location will occur before the process settles down to the solution. Additional steps are shown in Table 8.1.

TABLE 8.1

Steepest descent example convergence

x_0	x_1	x_2	x_3	x_4	x_5
1	0.7797	1.2354	1.1844	1.1214	1.0340
3	2.0246	2.3718	2.3144	2.2351	2.1255

8.3 The generalization of Newton's method

We assume that every function f_i is twice differentiable, at least in the neighborhood of the anticipated solution. The Jacobian matrix is formulated as

$$\underline{F}(\underline{x}) = \begin{bmatrix} \frac{\partial f_1}{\partial x_1} & \frac{\partial f_1}{\partial x_2} & \cdots & \frac{\partial f_1}{\partial x_n} \\ \frac{\partial f_2}{\partial x_1} & \frac{\partial f_2}{\partial x_2} & \cdots & \frac{\partial f_2}{\partial x_n} \\ \cdots & & & \\ \frac{\partial f_n}{\partial x_1} & \frac{\partial f_n}{\partial x_2} & \cdots & \frac{\partial f_n}{\partial x_n} \end{bmatrix}.$$

Using this in place of the derivative (hence it is called the tangent matrix), we generalize the "scalar" Newton method for this case as follows:

$$\underline{f}(\underline{x}_0) + \underline{F}(\underline{x}_0)(\underline{x}_1 - \underline{x}_0) = \underline{0}.$$

The solution of the inhomogeneous system of linear equations,

$$\underline{F}(\underline{x}_0)(\underline{x}_0 - \underline{x}_1) = \underline{f}(\underline{x}_0),$$

yields the new iterate. Hence, the Newton iteration scheme for the nonlinear system of equations is

$$\underline{F}(\underline{x}_i)(\underline{x}_i - \underline{x}_{i+1}) = \underline{f}(\underline{x}_i).$$

It is also possible to compute the inverse of the Jacobian matrix and iterate as

$$\underline{x}_{i+1} = \underline{x}_i - \underline{F}^{-1}(\underline{x}_i)\underline{f}(\underline{x}_i).$$

We need to press the issue again that the method is operational only if the inverse of the Jacobian matrix exists. The fact that one must evaluate this inverse at each iteration step renders this simple generalized Newton's method for nonlinear systems unattractive. The remedy for this is in the next section.

8.3.1 Computational example

We revisit the example from the last section and execute a step of Newton's method. The Jacobian matrix for this system is

$$F(\underline{x}) = \begin{bmatrix} 2x_1 & 1 \\ 1 & 2x_2 \end{bmatrix}.$$

The inverse of the Jacobian is

$$F^{-1}(\underline{x}) = \frac{1}{4x_1x_2 - 1} \begin{bmatrix} 2x_2 & -1 \\ -1 & 2x_1 \end{bmatrix}.$$

Now we are ready to proceed with the iteration. We start with a zero starting vector.

$$\underline{x}_0 = \begin{bmatrix} 0 \\ 0 \end{bmatrix}.$$

Based on the example in the past section, we compute

$$f(\underline{x}_0) = f(\underline{0}) = \begin{bmatrix} -3 \\ -5 \end{bmatrix},$$

and

$$F^{-1}(\underline{x}_0) = F^{-1}(\underline{0}) = \begin{bmatrix} 0 & 1 \\ 1 & 0 \end{bmatrix}.$$

With the above we may compute the next iterative solution as

$$\underline{x}_1 = \underline{x}_0 - \underline{F}^{-1}(\underline{x}_0)\underline{f}(\underline{x}_0) = \begin{bmatrix} 0 \\ 0 \end{bmatrix} - \begin{bmatrix} 0 & 1 \\ 1 & 0 \end{bmatrix}\begin{bmatrix} -3 \\ -5 \end{bmatrix} = \begin{bmatrix} 5 \\ 3 \end{bmatrix}.$$

Further steps of the process are summarized in Table 8.2.

TABLE 8.2
Nonlinear Newton's method

$\underline{x}_0$	$\underline{x}_1$	$\underline{x}_2$	$\underline{x}_3$	$\underline{x}_4$	$\underline{x}_5$
0	5	2.6102	1.5225	1.0966	1.0047
0	3	1.8983	1.8965	1.9790	1.9989

The successive iterates are approaching, albeit not monotonically, the analytic solution of

$$\underline{x} = \begin{bmatrix} 1 \\ 2 \end{bmatrix}.$$

An accuracy, acceptable by engineering standards, of better than 1% is obtained in step 5.

8.4 Quasi-Newton method

As mentioned before, the burden of recomputing the inverse of the Jacobian at each iteration step is significant, especially when there are a large number of equations involved. To alleviate this, a way of computing the inverse of a certain step based on the inverse of an earlier step is needed.

In order to do so, we replace the Jacobian matrix in the nonlinear Newton iteration with a matrix M_i. The iteration step with this matrix is then

$$\underline{x}_{i+1} = \underline{x}_i - M_i^{-1}\underline{f}(\underline{x}_i).$$

The process starts with

$$M_0 = F(\underline{x}_0),$$

and proceeds for $i = 1, 2, \ldots$ as

$$M_i = M_{i-1} + \underline{s}_i^T \frac{\underline{t}_i - M_{i-1}\underline{s}_i}{||\underline{s}_i||^2}.$$

Here,

$$\underline{s}_i = \underline{x}_i - \underline{x}_{i-1},$$

and

$$\underline{t}_i = \underline{f}(\underline{x}_i) - \underline{f}(\underline{x}_{i-1}).$$

The definition of the M_i matrix above, in essence, approximates the tangent matrix with a "secant" matrix. The special construction of the matrix enables us to find an efficient way to compute the inverse of M_i.

The method, often called the BFGS method, based on the initials of the four main contributors, Broyden [1], Fletcher [2], Goldfarb [3] and Shanno [5], is based on the Sherman–Morrison formula of linear algebra. The formula is valid for a nonsingular matrix M_{i-1}, for which the inverse M_{i-1}^{-1} is available. We update this matrix by a rank one matrix defined by vectors $\underline{u}, \underline{v}$ as

$$M_i = M_{i-1} + \underline{u}\underline{v}^T.$$

The inverse of the updated matrix may be computed from the inverse of the original matrix as

$$M_i^{-1} = M_{i-1}^{-1} - \frac{M_{i-1}^{-1}\underline{u}\underline{v}^T M_{i-1}^{-1}}{1 + \underline{v}^T M_{i-1}^{-1}\underline{u}},$$

assuming that

$$\underline{v}^T M_{i-1}^{-1} \underline{u} \neq -1.$$

Exploiting the construction of the matrix, we assign

$$\underline{u} = \frac{\underline{t}_i - M_{i-1}\underline{s}_i}{||\underline{s}_i||^2}$$

and

$$\underline{v} = \underline{s}_i.$$

We apply these as updates to the M_{i-1} matrix and after a considerable amount of algebraic work, one obtains

$$M_i^{-1} = M_{i-1}^{-1} + \frac{(\underline{s}_i - M_{i-1}^{-1}\underline{t}_i)\underline{s}_i^T M_{i-1}^{-1}}{\underline{s}_i^T M_{i-1}^{-1}\underline{t}_i}.$$

This formula is the more efficient in the computational aspect but less speedy in convergence than Newton's method.

8.4.1 Computational example

To illustrate this method we reuse the example of the preceding section.

$$x_1^2 + x_2 - 3 = 0,$$

and

$$x_1 + x_2^2 - 5 = 0.$$

We started with

$$\underline{x}_0 = \begin{bmatrix} 0 \\ 0 \end{bmatrix}$$

and computed the inverse of the Jacobian matrix at the starting solution as

$$F^{-1}(\underline{x}_0) = F^{-1}(\underline{0}) = \begin{bmatrix} 0 & 1 \\ 1 & 0 \end{bmatrix}.$$

With this the next iterative solution is

$$\underline{x}_1 = \begin{bmatrix} 5 \\ 3 \end{bmatrix}.$$

Now we are ready to proceed with the quasi-Newton method. We compute the two auxiliary vectors.

$$\underline{s}_1 = \underline{x}_1 - \underline{x}_0 = \begin{bmatrix} 5 \\ 3 \end{bmatrix}.$$

$$\underline{t}_1 = \underline{f}(\underline{x}_1) - \underline{f}(\underline{x}_0) = \begin{bmatrix} 25 \\ 9 \end{bmatrix} - \begin{bmatrix} -3 \\ -5 \end{bmatrix} = \begin{bmatrix} 28 \\ 14 \end{bmatrix}.$$

We now compute the components of M_1^{-1}.

$$M_0^{-1} = F^{-1}(\underline{x}_0) = \begin{bmatrix} 0 & 1 \\ 1 & 0 \end{bmatrix},$$

$$\underline{s}_1 - M_0^{-1}\underline{t}_1 = \begin{bmatrix} -9 \\ -25 \end{bmatrix},$$

$$\underline{s}_1^T M_0^{-1} = \begin{bmatrix} 3 & 5 \end{bmatrix},$$

$$\underline{s}_1^T M_0^{-1}\underline{t}_1 = \begin{bmatrix} 154 \end{bmatrix}.$$

With these, finally,

$$M_1^{-1} = M_0^{-1} + \frac{(\underline{s}_1 - M_0^{-1}\underline{t}_1)\underline{s}_1^T M_0^{-1}}{\underline{s}_1^T M_0^{-1}\underline{t}_1} = \begin{bmatrix} -0.1753 & 0,7078 \\ 0.5130 & -0.8117 \end{bmatrix}.$$

The next iterate is obtained as

$$\underline{x}_2 = \underline{x}_1 - M_1^{-1}\underline{f}(\underline{x}_1) = \begin{bmatrix} 3.013 \\ -2.520 \end{bmatrix}.$$

This example was meant to demonstrate the process, rather than the numerical nuances. It is hard to visualize how this last step approaches the exact solution, but these methods sometimes first take the process farther from the solution before zooming in on it.

Note that while the example functions used in this chapter are nonlinear polynomials, the techniques introduced are also applicable to mixed functions of trigonometric, exponential and other types, as long as the derivatives, required by the calculation of the Jacobian matrix, are computable.

8.5 Nonlinear static analysis application

A very important application example of these techniques is in the area of nonlinear structural analysis. In most real-life environments, structural deformations are nonlinear. It is possible that the structure undergoes a large deformation, not linearly related to the increase in the applied load. It is also possible that the structural material is nonlinearly elastic, or even plastic. For more on the engineering aspects of these phenomena, see [4].

In both cases, the problem is presented as

$$K(\underline{u})\underline{u} = \underline{f},$$

where the matrix $K(u)$, representing the structure's stiffness characteristics, is called the stiffness matrix and u is the deformation of the structure. Note that the stiffness matrix is dependent on the deformation. The right-hand side contains a vector representing the loads applied to the structure.

The size of the matrix is related to the finite element discretization of the structure. It can have millions of rows and columns in today's engineering practice, hence efficient solution techniques are a necessity. Both the generalized Newton method and the quasi-Newton method are applicable and used in the industry to solve these problems.

True to the nonlinear nature of the problem, there is a load imbalance between the external load $\underline{f}$ and the internally absorbed load

$$\Delta f_i = f - K(\underline{u}_i)\underline{u}_i.$$

The iteration

$$\underline{u}_{i+1} = \underline{u}_i + K_i^{-1}\Delta f_i$$

proceeds until the load imbalance Δf_i disappears, which is when the structure has reached an equilibrium.

In practical applications, a rank two update version of the BFGS process is executed. Two regular Newton iterations are executed and a BFGS update based on the two steps is formulated. We introduce the notation of

$$\gamma = \Delta f_{i-1} - \Delta f_i,$$

which are the force imbalances at the two steps, and

$$\delta = u_i - u_{i-1}.$$

The approximate updated stiffness matrix based on the last two points and the just introduced quantities is computed as

$$K_{i+1} = K_i + \frac{\gamma\gamma^T}{\gamma^T\delta} - \frac{K_i\delta\delta^T K_i}{\delta^T K_i\delta}.$$

This is a rank two update of a matrix. The related BFGS update [4] is

$$K_{i+1}^{-1} = K_i^{-1} + (1 + \frac{\gamma^T K_i^{-1}\gamma}{\gamma^T\delta})\frac{\delta\delta^T}{\delta^T\gamma} - \frac{\delta\gamma^T K_i^{-1} + K_i^{-1}\gamma\delta^T}{\gamma^T\delta}.$$

This is usually reformulated for computational purposes. Nevertheless, it directly produces the inverse for the next approximate stiffness from the last inverse and from information in the last two steps, hence it is a secant-type approximation.

References

[1] Broyden, C. G.; The convergence of a class of double rank minimization algorithms, *J. Inst. Math. Appl.*, Vol. 6, pp. 76–90, 222-231, 1970.

[2] Fletcher, R.; A new approach to variable metric algorithms, *Computer Journal*, Vol. 13, pp. 317–322, 1970.

[3] Goldfarb, D.; A family of variable metric methods derived by variational means, *Math. Comp.*, Vol. 24, pp. 23–26, 1970.

[4] Komzsik, L.; *Computational Techniques of Finite Element Analysis*, CRC Press, Boca Raton, FL, 2005.

[5] Shanno, D. F.; Conditioning of quasi-Newton methods for function minimization, *Math. Comp.*, Vol. 24, pp. 647–656, 1970.

[6] Ortega, J. M. and Rheinboldt, W. C.; *Iterative Solution of Nonlinear Equations in Several Variables*, SIAM, Philadelphia, 2000.

9

Iterative solution of linear systems

This topic has also been the subject of historical interest for many centuries. Gauss had already researched the topic by the early 1800s (published posthumously in 1903 [1]) and some of the methods bear his name in recognition of his laying the foundation. Jacobi [6] also published in the area over 150 years ago, and Seidel's and Krylov's [9] contributions are rather old, too. Similarly, Ritz laid the foundation for some of this chapter's topics in [10], albeit not formally describing the technique now called the Ritz–Galerkin procedure.

The most significant result came in the middle of the last century: the method of conjugate gradients by Hestenes and Stiefel [3]. The method and its variants have found their way into many aspects of engineering applications and are still the subject of active research.

9.1 Iterative solution of linear systems

The problem of solving the system of linear equations,

$$a_{11}x_1 + a_{12}x_2 + \ldots + a_{1n}x_n = b_1,$$

$$a_{21}x_1 + a_{22}x_2 + \ldots + a_{2n}x_n = b_2,$$

$$\ldots$$

$$a_{n1}x_1 + a_{n2}x_2 + \ldots + a_{nn}x_n = b_n,$$

is discussed by taking advantage of the convenience of matrix notation. The problem is then

$$A\underline{x} = \underline{b},$$

with

$$\underline{b} \neq \underline{0},$$

and

$$det(A) \neq 0.$$

The iterative solution of this system may be found by a splitting of the matrix into two additive components as

$$A = B - C,$$

with the assumption of

$$det(B) \neq 0.$$

Furthermore, we desire that B^{-1} be easy to compute. The system of equations may be reordered in terms of these components as

$$B\underline{x} = C\underline{x} + \underline{b}.$$

This enables a conceptually very simple iteration scheme.

$$\underline{x} = B^{-1}C\underline{x} + B^{-1}\underline{b}.$$

Computing the matrix

$$M = B^{-1}C,$$

and a vector of

$$\underline{c} = B^{-1}\underline{b},$$

one obtains the iterative sequence of

$$\underline{x}_{i+1} = M\underline{x}_i + \underline{c}.$$

The process may be started with $\underline{x}_0 = \underline{0}$. Furthermore, let us denote the exact solution with $\underline{x}$ and let the error of the iteration steps be represented by

$$\underline{e}_1 = \underline{x}_1 - \underline{x} = M(\underline{x}_0 - \underline{x}) = M\underline{e}_0,$$
$$\underline{e}_2 = \underline{x}_2 - \underline{x} = M(\underline{x}_1 - \underline{x}) = M^2\underline{e}_0,$$

and

$$\underline{e}_{i+1} = \underline{x}_{i+1} - \underline{x} = M(\underline{x}_i - \underline{x}) = M^{i+1}\underline{e}_0.$$

The sequence of approximations will converge to the exact solution as long as

$$lim_{i \to \infty} M^i = (\underline{0}),$$

which is satisfied if

$$||M|| = ||B^{-1}C|| \leq 1.$$

The above is a necessary condition for the convergence of an iterative solution of a system based on a particular splitting.

9.2 Splitting methods

There are several well-known classical strategies for the splitting of the system matrix.

9.2.1 Jacobi method

One of the simplest techniques, well known to engineers, is the Jacobi method [6]. The B matrix is simply defined as

$$B = diag(A(j,j)), j = 1, 2, \ldots, n,$$

a diagonal matrix. The iteration scheme of

$$\underline{x}_{i+1} = B^{-1}C\underline{x}_i + B^{-1}\underline{b}$$

becomes the following term wise formula for every $j = 1, 2, \ldots, n$.

$$x_{i+1}(j) = \frac{1}{A(j,j)}(b(j) - \sum_{k=1}^{j-1} A(j,k)x_i(k) - \sum_{k-j+1}^{n} A(j,k)x_i(k)).$$

The convergence condition is

$$||B^{-1}C|| = max_{1 \leq j \leq n} \sum_{\substack{k \neq j}} |\frac{A(j,k)}{A(j,j)}| < 1.$$

It follows that

$$\sum_{k \neq j} |A(j,k)| < |A(j,j)|; j = 1, \ldots, n$$

is required, which means the matrix A must be diagonally dominant for the Jacobi method to converge.

9.2.2 Gauss–Seidel method

An improvement is proposed by Seidel and presented in the Gauss–Seidel method by using the just computed new iterative values on the right-hand side [12].

$$x_{i+1}(j) = \frac{1}{A(j,j)}(b(j) - \sum_{k=1}^{j-1} A(j,k)x_{i+1}(k) - \sum_{k=j+1}^{n} A(j,k)x_i(k)).$$

The method may be written in matrix form as

$$\underline{x}_{i+1} = (D + L)^{-1}(\underline{b} - U\underline{x}_i).$$

Here,

$$A = L + D + U,$$

where L, D and U are the strictly lower triangular, diagonal and strictly upper triangular parts of the A matrix, respectively. This is the basis for a computationally more useful form. Expand as

$$\underline{x}_{i+1} = (D + L)^{-1}\underline{b} - (D + L)^{-1}U\underline{x}_i.$$

Adding and subtracting

$$(D + L)^{-1}(D + L)\underline{x}_i,$$

and substituting

$$\underline{r}_i = \underline{b} - A\underline{x}_i = \underline{b} - (L + D + U)\underline{x}_i,$$

one gets

$$\underline{x}_{i+1} = \underline{x}_i + (D + L)^{-1}\underline{r}_i.$$

We seemingly lost the U partition of the matrix, but it is still hidden in the residual $\underline{r}_i$. Starting from an initial estimated solution of $\underline{x}_0$ and computing the last two equations for $i = 1, 2 \ldots$ result in an approximate solution. The convergence condition is the same as that of the Jacobi method; however, the rate of convergence is usually about twice as fast.

9.2.3   Successive overrelaxation method

The Gauss–Seidel method may be extended by applying a relaxation factor, resulting in the SOR, successive overrelaxation technique [15]. The Gauss–Seidel matrix form, introduced above, is extended with the relaxation factor ω as

$$\underline{x}_{i+1} = (D + \omega L)^{-1}(-\omega U + (1 - \omega)D)\underline{x}_i + \omega(D + \omega L)^{-1}\underline{b}.$$

One can see that the value of

$$\omega = 1$$

results in the Gauss–Seidel method. Values higher than one technically result in overrelaxation and those lower than one, in underrelaxation. However, in common practice today, overrelaxation is defined by any value other than one.

This method, depending on the selection of the relaxation factor, may be an order of magnitude faster than the Gauss–Seidel method. In fact, for a given matrix, an optimal relaxation value may be theoretically computed as

$$\omega_{optimum} = \frac{2}{1 + \sqrt{1 - \rho^2}},$$

where ρ is the spectral radius of the matrix. As computing the spectral radius exactly is a more expensive proposition than solving the system of equations directly, this optimum is mostly of theoretical interest. For special matrices where the spectral radius may be approximated, this formula is used.

It is proven in [7] that the limits of the relaxation value are

$$0 < \omega < 2.$$

Convergence will not occur outside of that region. On the other hand, if the matrix is symmetric and positive definite, the method is guaranteed to converge for any relaxation value between these limits. The actual value of the relaxation parameter heavily influences the rate of convergence.

9.2.4 Computational formulation of splitting methods

Finally, any particular splitting technique, placing a focus on the error of iteration in each step, may be rearranged into a computationally convenient form. In order to do so the

$$A = B - C$$

splitting equation is used in the

$$C = B - A$$

form. Similarly the iterative scheme of

$$B\underline{x}_{i+1} = C\underline{x}_i + \underline{b}$$

is rewritten as

$$B\underline{x}_{i+1} = (B - A)\underline{x}_i + \underline{b} = B\underline{x}_i + \underline{b} - A\underline{x}_i.$$

We introduce the residual vector

$$\underline{r}_i = \underline{b} - A\underline{x}_i,$$

and a difference vector

$$\underline{\Delta x}_i = \underline{x}_{i+1} - \underline{x}_i.$$

The convenient computational form of

$$B\underline{\Delta x}_i = \underline{r}_i$$

is the basis of an iteration scheme in terms of the residual vector as follows. At step i, first compute the residual

$$\underline{r}_i = \underline{b} - A\underline{x}_i,$$

then the adjustment of the solution

$$\underline{\Delta x}_i = B^{-1}\underline{r}_i,$$

followed by the new iterate as

$$\underline{x}_{i+1} = \underline{x}_i + \underline{\Delta x}_i.$$

The convergence condition is still based on

$$||B^{-1}C|| = ||B^{-1}(B - A)|| = ||I - B^{-1}A|| < 1.$$

The residual of the above process is readily usable for a stopping criterion.

9.3   Ritz–Galerkin method

The most useful iterative techniques for solution of linear systems in engineering practice today are various conjugate gradient methods. These methods are based on the Ritz–Galerkin principle that proposes to select such iterative solution vectors $\underline{x}_i$ for which the residual is orthogonal to a Krylov subspace generated by A and the initial residual $\underline{r}_0$.

This Krylov subspace K^i is defined by the sequence of vectors generated as

$$\underline{k}_1, \underline{k}_2, \ldots, \underline{k}_i,$$

where

$$\underline{k}_i = A^{i-1}\underline{r}_0.$$

These vectors do not describe the subspace well because they are not orthogonal. Let us assume for now that we have a set of orthonormal basis vectors spanning the same Krylov subspace in

$$V_i = \begin{bmatrix} \underline{v}_1 \ \underline{v}_2 \ \cdots \ \underline{v}_i \end{bmatrix}.$$

The Ritz–Galerkin principle in terms of this yet unknown orthonormal spanning set of the Krylov subspace is described as

$$V_i^T \underline{r}_i = 0,$$

or

$$V_i^T (\underline{b} - A\underline{x}_i) = 0.$$

To generate these vectors we start from

$$\underline{x}_0 = \underline{0},$$

which results in

$$\underline{r}_0 = \underline{b}.$$

We compute the first v vector by normalizing as

$$\underline{v}_1 = \frac{\underline{r}_0}{||\underline{r}_0||}.$$

The Ritz–Galerkin condition may be expanded and written as

$$V_i^T \underline{b} = V_i^T A \underline{x}_i.$$

By the orthogonality of the V basis and the definition of the $\underline{v}_1$ vector,

$$V_i^T \underline{b} = ||r_0||\underline{e}_1.$$

Selecting $\underline{x}_i$ from the Krylov subspace as

$$\underline{x}_i = V_i \underline{y}_i$$

we obtain

$$V_i^T A V_i \underline{y}_i = ||\underline{r}_0||\underline{e}_1.$$

The form on the left-hand side is, in essence, a projection of the A matrix into the Krylov subspace spanned by V. In the case of a symmetric A matrix this projection produces a tridiagonal matrix

$$V_i^T A V_i = T_i.$$

The process of systematically generating the vectors of V is the Lanczos method, whose virtues will be extolled at length in the next chapter. The solution for the next iterate will be generously accelerated by the tridiagonal matrix form, specifically,

$$T_i \underline{y}_i = \|\underline{r}_0\| \underline{e}_1.$$

Clearly the solution for the new iterate is

$$\underline{y}_i = T_i^{-1} \|\underline{r}_0\| \underline{e}_1,$$

and

$$\underline{x}_i = V_i \underline{y}_i.$$

The inverse of the tridiagonal matrix is not formally evaluated and in fact the execution of this step leads to the conjugate gradient method.

9.4   Conjugate gradient method

The exploitation of the tridiagonal form in the case of positive definite matrices yields the method of conjugate gradients [3]. Positive definiteness implies that the tridiagonal matrix may be factored without pivoting into

$$T_i = L_i U_i.$$

Here the factor matrices are a lower bidiagonal matrix of

$$L_i = \begin{bmatrix} d_1 & & & \\ l_1 & d_2 & & \\ & \ldots & \ldots & \\ & & l_{i-1} & d_i \end{bmatrix},$$

and a unit upper bidiagonal matrix of

$$U_i = \begin{bmatrix} 1 & u_1 & & \\ & \ldots & \ldots & \\ & & 1 & u_{i-1} \\ & & & 1 \end{bmatrix}.$$

We use this factorization in order to produce the inverse

$$\underline{y}_i = U_i^{-1} L_i^{-1} (\|\underline{r}_0\| \underline{e}_1).$$

The following clever algorithm, invented by Hestenes and Stiefel [3], avoids the explicit computations of the terms of the tridiagonal matrix, as well as its factorization. Instead the following process is executed. Initialize as

$$\underline{x}_0 = \underline{0}, \underline{r}_0 = \underline{b}, \underline{p}_0 = \underline{r}_0,$$

where p_0 is a starting search direction vector. The process iterates through the following five steps for $i = 1, 2, 3, \ldots$ until convergence.

1. Compute the length of the step

$$\alpha_i = \frac{r_{i-1}^T r_{i-1}}{p_{i-1}^T A p_{i-1}}.$$

2. Compute the next approximate solution

$$x_i = x_{i-1} + \alpha_i p_{i-1}.$$

3. Compute the residual of the new approximate solution

$$r_i = r_{i-1} - \alpha_i A p_{i-1}.$$

4. Compute the relative improvement of this step

$$\beta_i = \frac{r_i^T r_i}{r_{i-1}^T r_{i-1}}.$$

5. Compute the new search direction

$$p_i = r_i + \beta_i p_{i-1}.$$

The process may be stopped when the norm of the residual computed in step 3,

$$\|r_i\|,$$

is less than a certain threshold.

The coefficients generated during the conjugate gradient steps are related to the tridiagonal matrix, which was not built explicitly. The coefficients are

$$T_i = \begin{bmatrix} \ddots & & \\ \cdot & -\frac{\beta_{i-1}}{\alpha_{i-1}} & \cdot \\ \cdot & \frac{1}{\alpha_i} + \frac{\beta_{i-1}}{\alpha_{i-1}} & \cdot \\ \cdot & -\frac{1}{\alpha_i} & \cdot \\ & & \ddots \end{bmatrix}$$

One can proceed now and actually compute the eigenvalues of the matrix of the system, but that topic and more efficient methods of doing it will be addressed in the next chapter.

9.4.1   Computational example

We will execute the conjugate gradient process for the system

$$A\underline{x} = \begin{bmatrix} 20 & 1 & 0 \\ 1 & 20 & -1 \\ 0 & -1 & 20 \end{bmatrix} \begin{bmatrix} x_1 \\ x_2 \\ x_3 \end{bmatrix} = \begin{bmatrix} 21 \\ 20 \\ 19 \end{bmatrix}$$

until we obtain a residual norm of less than 0.1.

Initialization:

$$\underline{r}_0 = \begin{bmatrix} 21 \\ 20 \\ 19 \end{bmatrix} = \underline{p}_0.$$

Step $i = 1$:

1. Compute the length of the step

$$\alpha_1 = \frac{\underline{r}_0^T \underline{r}_0}{\underline{p}_0^T A \underline{p}_0} = \frac{\begin{bmatrix} 21 \\ 20 \\ 19 \end{bmatrix}^T \begin{bmatrix} 21 \\ 20 \\ 19 \end{bmatrix}}{\begin{bmatrix} 21 \\ 20 \\ 19 \end{bmatrix}^T A \begin{bmatrix} 21 \\ 20 \\ 19 \end{bmatrix}} = 0.049834.$$

2. Compute the next approximate solution

$$\underline{x}_1 = \underline{x}_0 + \alpha_1 \underline{p}_0 = \begin{bmatrix} 0 \\ 0 \\ 0 \end{bmatrix} + 0.049834 \begin{bmatrix} 21 \\ 20 \\ 19 \end{bmatrix} = \begin{bmatrix} 1.04652 \\ 0.99668 \\ 0.94685 \end{bmatrix}.$$

3. Compute the residual of the new approximation

$$\underline{r}_1 = \underline{r}_0 - \alpha_1 A \underline{p}_0 = \begin{bmatrix} 21 \\ 20 \\ 19 \end{bmatrix} - \alpha_1 A \begin{bmatrix} 21 \\ 20 \\ 19 \end{bmatrix} = \begin{bmatrix} -0.92703 \\ 0.03333 \\ 1.05970 \end{bmatrix}.$$

4. Compute the relative improvement of this step

$$\beta_1 = \frac{\underline{r}_1^T \underline{r}_1}{\underline{r}_0^T \underline{r}_0} = \frac{\begin{bmatrix} -0.92703 \\ 0.03333 \\ 1.05970 \end{bmatrix}^T \begin{bmatrix} -0.92703 \\ 0.03333 \\ 1.05970 \end{bmatrix}}{\begin{bmatrix} 21 \\ 20 \\ 19 \end{bmatrix}^T \begin{bmatrix} 21 \\ 20 \\ 19 \end{bmatrix}} = 0.00165.$$

5. Compute the new search direction

$$\underline{p}_1 = \underline{r}_1 + \beta_1 \underline{p}_0 = \begin{bmatrix} -0.92703 \\ 0.03333 \\ 1.05970 \end{bmatrix} + \beta_1 \begin{bmatrix} 21 \\ 20 \\ 19 \end{bmatrix} = \begin{bmatrix} -0.89238 \\ -0.00033 \\ 1.09110 \end{bmatrix}.$$

Step $i = 2$:

1. Compute the length of the step

$$\alpha_2 = \frac{\underline{r}_1^T \underline{r}_1}{\underline{p}_1^T A \underline{p}_1} = 0.049916.$$

2. Compute the next approximate solution

$$\underline{x}_2 = \underline{x}_1 + \alpha_2 \underline{p}_1 = \begin{bmatrix} 1.00197 \\ 0.99667 \\ 1.00131 \end{bmatrix}.$$

3. The residual computation yields

$$\underline{r}_2 = \underline{r}_1 - \alpha_2 A \underline{p}_1 = \begin{bmatrix} -0.03614 \\ 0.06600 \\ -0.02954 \end{bmatrix}.$$

The residual norm is

$$||\underline{r}_2|| = \sqrt{\underline{r}_2^T \underline{r}_2} = 0.0808.$$

This satisfies our accuracy requirement for this example. The approximate solution is appropriately close to the exact solution of

$$\underline{x} = \begin{bmatrix} 1 \\ 1 \\ 1 \end{bmatrix}.$$

9.5 Preconditioning techniques

To accelerate the convergence of the iteration, a preconditioning step is often applied. This entails premultiplying the system by the inverse of a positive definite preconditioning matrix P.

$$P^{-1} A \underline{x} = P^{-1} \underline{b}.$$

The intention is, of course, to have the

$$\overline{A} = P^{-1}A$$

matrix produce better convergence. One way to apply the preconditioner is in its factored form

$$P = LL^T$$

and applied as

$$L^{-1}AL^{-1,T}\underline{y} = L^{-1}\underline{b}.$$

Then the original solution is recovered as

$$\underline{x} = L^{-1,T}\underline{y}.$$

The conjugate gradient method, however, allows for a more elegant solution. Let us define a P inner product as

$$(x,y)_P = (x, Py).$$

While the $P^{-1}A$ matrix is not symmetric, even if A is symmetric, it is symmetric with respect to the P inner product.

$$(P^{-1}Ax, y)_P = (P^{-1}Ax, Py) = (Ax, y) = (x, Ay) = (x, P^{-1}Ay)_P.$$

If we can use the conjugate gradient procedure in the P inner product, we minimize

$$(\underline{r}_i, P^{-1}A\underline{r}_i)_P = (\underline{r}_i, A\underline{r}_i).$$

The outcome of this is that the preconditioned conjugate gradient process will still minimize the error of the original system, however, now in relation to a Krylov subspace generated by $P^{-1}A$. This makes the necessary modification to the preconditioned conjugate gradient algorithm minimal. The process needs the introduction of a step to apply the preconditioner to the residual whenever it is computed as

$$P\underline{t}_i = \underline{r}_i.$$

In addition, the computation of the new search direction is executed in terms of this temporary $\underline{t}$ vector.

$$\underline{p}_i = \underline{t}_i + \beta_i\underline{p}_{i-1}.$$

The preconditioner P may also be factored in this context, facilitating a speedy application via forward-backward substitution.

9.5.1 Approximate inverse

As far as the actual content of the preconditioner is concerned, most successful techniques are based on the preconditioner being an approximate inverse of the A matrix. Such may be obtained by incomplete factorizations [8].

The concept is to execute a factorization based on the Gaussian elimination [1]

$$A = LU,$$

but compute the fill-in terms only at certain positions. One simple way to define this restriction is by enforcing that

$$L(i, j), U(i, j) = 0, \; if \; A(i, j) = 0.$$

This means retaining the sparsity pattern of the original matrix in the factors. To compute these incomplete factors is much less expensive than the complete factors. The computed incomplete factors $\overline{L}, \overline{U}$ only approximate the matrix

$$\overline{L}\,\overline{U} \approx A.$$

Hence if they are used as factors of the preconditioner

$$P = \overline{L}\,\overline{U},$$

the preconditioner P^{-1} will represent an approximate inverse of the matrix.

There are other methods of restricting the computation of the fill-in terms of the Gaussian elimination; for example, a band profile is often used. In this case fill-ins are ignored outside the band:

$$L(i, j) = 0, \; if \; j > i + b,$$

where b is a predefined bandwidth. Then there are methods of suppressing fill-ins based on their numerical value as opposed to location. In this case it is of course necessary to compute the fill-in terms, losing some of the computational efficiency advantage. Still, there is significant saving when the incomplete factor is applied repeatedly in the iteration scheme, as it contains fewer terms and is numerically more accurate. These methods are successful in certain applications and less so in others.

The same approach may also be used in connection with Cholesky decomposition.

$$A \approx \overline{C}\,\overline{C}^{T},$$

and

$$P = \overline{C}\,\overline{C}^{T}.$$

The use of incomplete Cholesky preconditioners for symmetric problems in connection with the preconditioned conjugate gradient method is widespread.

9.6   Biconjugate gradient method

So far we have focused on symmetric matrices. For the unsymmetric case a generalization of the conjugate gradient method exists, known as the biconjugate gradient method. Its underlying principle is still the Ritz–Galerkin, sometimes called Petrov–Galerkin, approach of orthogonalizing the successive residuals to the already gathered Krylov subspace. Due to the unsymmetric nature of the A matrix there are two distinct Krylov bases, U_i, V_i, to perform the tridiagonal reduction.

$$U_i^T A V_i = T_i,$$

with

$$U_i^T V_i = I.$$

The biorthogonal Lanczos method discussed in detail in the next chapter is readily available to accomplish this operation and the tridiagonal matrix in general is unsymmetric. The Ritz–Galerkin condition in this case becomes

$$U_i^T A V_i \underline{y} - U_i^T \underline{b} = \underline{0}.$$

The solution will again be based on

$$T_i \underline{y}_i = \|\underline{r}_0\| \underline{e}_1.$$

As in the symmetric conjugate gradient scheme, the computation of the actual tridiagonal reduction is skipped and a biconjugate algorithm is applied. The steps of this algorithm are as follows. Initialize as

$$\underline{x}_0 = \underline{0}, \ \underline{r}_0 = \underline{b}, \ \underline{p}_0 = \underline{r}_0, \ \underline{q}_0 = \underline{s}_0 = random,$$

where $\underline{q}_i$, $\underline{s}_i$ are counterparts of $\underline{p}_i$, $\underline{r}_i$ to enforce the biorthogonality condition. The biconjugate gradient process iterates until convergence through the same five steps as the symmetric conjugate gradients, but in steps 3 and 5, the orthogonal counterparts are also computed.

1. Compute the length of the step

$$\alpha_i = \frac{\underline{s}_{i-1}^T \underline{r}_{i-1}}{\underline{q}_{i-1}^T A \underline{p}_{i-1}}.$$

2. Compute the next approximate solution

$$\underline{x}_i = \underline{x}_{i-1} + \alpha_i \underline{p}_{i-1}.$$

3. Compute the residual of the new approximation

$$\underline{r}_i = \underline{r}_{i-1} - \alpha_i A \underline{p}_{i-1},$$

$$\underline{s}_i = \underline{r}_{i-1} - \alpha_i A^T \underline{q}_{i-1}.$$

4. Compute the relative improvement of this step

$$\beta_i = \frac{\underline{s}_i^T \underline{r}_i}{\underline{s}_{i-1}^T \underline{r}_{i-1}}.$$

5. Compute the new search direction and its counterpart

$$\underline{p}_i = \underline{r}_i + \beta_i \underline{p}_{i-1},$$

$$\underline{q}_i = \underline{s}_i + \beta_i \underline{q}_{i-1}.$$

The process may be stopped when both residual norms,

$$\|\underline{r}_i\|, \|\underline{s}_i\|,$$

computed in step 3, are less than a certain threshold.

The biconjugate gradient method, due to the underlying biorthogonal Lanczos reduction, suffers from the possibility of a breakdown. It is possible that the

$$\underline{u}_i^T \underline{v}_i$$

inner product is zero, even though neither vector is zero. This scenario manifests itself in the biconjugate gradient algorithm's fourth step with

$$\underline{s}_{i-1}^T \underline{r}_{i-1}$$

becoming zero and resulting in a division by zero. There are remedies for such occurrences, such as the look-ahead Lanczos [11], which overcomes the difficulty by creating a block-step consisting of two single steps, hence introducing a "bulge" in the otherwise tridiagonal matrix.

Another way to deal with this problem is to restart the Krylov subspace building process when the breakdown condition approaches. This technique is widely used in the industrial implementations of the Lanczos method and will be discussed in the next chapter.

9.7   Least squares systems

So far we have focused on square systems of equations. There are cases when the number of rows of the system matrix is less than the number of columns, the so-called underdetermined system. Then there are cases when the opposite is true: there are more rows than columns, i.e., there are more equations than unknowns. This class of overdetermined problems of

$$A\underline{x} = \underline{b},$$

in which the matrix A has n rows and m columns, and

$$n > m,$$

is the subject of this chapter. We will assume that the matrix has full column (m) rank.

Such a system has no exact solution, so an approximate solution is sought. It is rather intuitive to find an approximate solution using the earlier discussed least squares principle. The principle applied to this problem is

$$||A\underline{x} - \underline{b}|| = min,$$

where the use of the Euclidean norm asserts the square aspect of the least squares, since the above is equivalent to

$$(A\underline{x} - \underline{b})^T (A\underline{x} - \underline{b}) = min.$$

From executing the operations and differentiating with respect to $\underline{x}$, it follows that the least squares minimal solution satisfies

$$A^T (A\underline{x} - \underline{b}) = 0.$$

This is the so-called normal equation:

$$A^T A\underline{x} = A^T \underline{b}.$$

Explicitly executing this operation is numerically disadvantageous and computationally inefficient. A better way to solve this problem is by executing a factorization of

$$A = QR,$$

where Q is an orthogonal $n \times m$ matrix, i.e., $Q^T Q = I$. The R matrix is upper triangular of size m. The details of this factorization are addressed in the next section. Exploiting the characteristics of this factorization

$$Q^T A = R$$

and writing the overdetermined system as

$$Q^T A \underline{x} = Q^T \underline{b}$$

yields

$$R\underline{x} = Q^T \underline{b}.$$

Due to the upper triangular nature of the R matrix, the solution of this is a convenient and easily executable back substitution.

9.7.1   QR factorization

The factorization of

$$A = QR$$

may be accomplished with the systematic orthogonalization of the columns of the A matrix via the Gram–Schmidt orthogonalization scheme already mentioned. Here the dimensions of the matrices are: A and Q have n rows and m columns and R is of size m rows by m columns, where only the upper triangular matrix contains the nonzero terms [2].

Let us denote the columns of A with $\underline{a}_k$ and the columns of Q as $\underline{q}_k$. Then the factorization is represented by

$$\underline{a}_k = \sum_{i=1}^{k} r_{ik} \underline{q}_i, k = 1, 2, \ldots, m.$$

Here r_{ik} are the terms of the R matrix. As we have assumed that the A matrix has the full rank of m, all the r_{kk} terms are nonzero and we can rewrite

$$\underline{q}_k = \frac{1}{r_{kk}} (\underline{a}_k - \sum_{i=1}^{k-1} (\underline{q}_i^T \underline{a}_k) \underline{q}_i),$$

which is the well-known Gram–Schmidt orthogonalization scheme. Introducing

$$r_{ik} = \underline{q}_i^T \underline{a}_k$$

and

$$\underline{z}_k = r_{kk} \underline{q}_k$$

brings the computational form

$$\underline{z}_k = \underline{a}_k - \sum_{i=1}^{k-1} r_{ik}\underline{q}_i$$

and

$$r_{kk} = \sqrt{\underline{z}_k^T \underline{z}_k}.$$

The process starts with

$$r_{11} = \sqrt{\underline{a}_1^T \underline{a}_1}$$

and proceeds to $k = 2, \ldots, m$.

The process may be also executed with the help of Householder reflections [4]. That is the preferred method when the problem sizes are very large and the matrix A is square. Note that a square matrix may still be a least squares system if the column rank of the matrix is less than the number of columns.

9.7.2 Computational example

We consider the overdetermined system of

$$A\underline{x} = \underline{b},$$

with a rectangular matrix

$$A = \begin{bmatrix} 1 & 1 \\ 1 & 1 \\ 0 & 2 \end{bmatrix}$$

and right-hand side

$$\underline{b} = \begin{bmatrix} 3 \\ 3 \\ 4 \end{bmatrix}.$$

From

$$A^T A = \begin{bmatrix} 2 & 2 \\ 2 & 6 \end{bmatrix},$$

it is clear that the columns are not orthogonal. We first execute the QR factorization of the matrix.

Step $k = 1$:

$$r_{11}^2 = \begin{bmatrix} 1 \\ 1 \\ 0 \end{bmatrix}^T \begin{bmatrix} 1 \\ 1 \\ 0 \end{bmatrix} = 2,$$

from which

$$\underline{q}_1 = \frac{1}{r_{11}} \underline{a}_1 = \begin{bmatrix} \frac{1}{\sqrt{2}} \\ \frac{1}{\sqrt{2}} \\ 0 \end{bmatrix}.$$

Step $k = 2$:

$$r_{12} = \underline{q}_1^T \underline{a}_2 = \begin{bmatrix} \frac{1}{\sqrt{2}} \\ \frac{1}{\sqrt{2}} \\ 0 \end{bmatrix}^T \begin{bmatrix} 1 \\ 1 \\ 2 \end{bmatrix} = \sqrt{2}.$$

$$\underline{z}_2 = \underline{a}_2 - \sqrt{2}\underline{q}_1 = \begin{bmatrix} 1 \\ 1 \\ 2 \end{bmatrix} - \sqrt{2} \begin{bmatrix} \frac{1}{\sqrt{2}} \\ \frac{1}{\sqrt{2}} \\ 0 \end{bmatrix} = \begin{bmatrix} 0 \\ 0 \\ 2 \end{bmatrix}.$$

Finally,

$$r_{22}^2 = \begin{bmatrix} 0 \\ 0 \\ 2 \end{bmatrix}^T \begin{bmatrix} 0 \\ 0 \\ 2 \end{bmatrix} = 4,$$

and

$$\underline{q}_2 = \begin{bmatrix} 0 \\ 0 \\ 1 \end{bmatrix}.$$

This step is representative of all the following steps in the case of a problem with more columns.

Assembling the results:

From the computed components:

$$Q = \begin{bmatrix} \frac{1}{\sqrt{2}} & 0 \\ \frac{1}{\sqrt{2}} & 0 \\ 0 & 1 \end{bmatrix},$$

and

$$R = \begin{bmatrix} \sqrt{2} & \sqrt{2} \\ 0 & 2 \end{bmatrix}.$$

Verification step:

$$Q^T Q = \begin{bmatrix} \frac{1}{\sqrt{2}} & 0 \\ \frac{1}{\sqrt{2}} & 0 \\ 0 & 1 \end{bmatrix}^T \begin{bmatrix} \frac{1}{\sqrt{2}} & 0 \\ \frac{1}{\sqrt{2}} & 0 \\ 0 & 1 \end{bmatrix} = \begin{bmatrix} 1 & 0 \\ 0 & 1 \end{bmatrix} = I,$$

and

$$QR = \begin{bmatrix} \frac{1}{\sqrt{2}} & 0 \\ \frac{1}{\sqrt{2}} & 0 \\ 0 & 1 \end{bmatrix} \begin{bmatrix} \sqrt{2} & \sqrt{2} \\ 0 & 2 \end{bmatrix} = \begin{bmatrix} 1 & 1 \\ 1 & 1 \\ 0 & 2 \end{bmatrix} = A.$$

Now we turn to the solution of the least squares system.

Multiplication of the right-hand side with the Q^T matrix results in

$$Q^T \underline{b} = \begin{bmatrix} \frac{1}{\sqrt{2}} & 0 \\ \frac{1}{\sqrt{2}} & 0 \\ 0 & 1 \end{bmatrix}^T \begin{bmatrix} 3 \\ 3 \\ 4 \end{bmatrix} = \begin{bmatrix} 3\sqrt{2} \\ 4 \end{bmatrix}.$$

The back substitution of

$$R\underline{x} = \begin{bmatrix} \sqrt{2} & \sqrt{2} \\ 0 & 2 \end{bmatrix} \begin{bmatrix} x_1 \\ x_2 \end{bmatrix} = \begin{bmatrix} 3\sqrt{2} \\ 4 \end{bmatrix}$$

produces the least squares solution of the system:

$$\underline{x} = \begin{bmatrix} 1 \\ 2 \end{bmatrix}.$$

Careful observation shows that this is really an exact solution of the system as the first two equations were identical, so the problem really was a square system only. Nevertheless, it was a simple way to demonstrate the technology.

9.8   The minimum residual approach

There is another class of methods that minimizes the norm of the residual at each iteration; hence they are called minimum norm residual methods. The method gathers approximate solutions that minimize the

$$\|A\underline{x}_i - \underline{b}\|$$

norm at every step. In the generic nonsymmetric A matrix case, the Krylov space for such a set of vectors $\underline{v}_i$ projects

$$AV_i = V_i H_i,$$

where H is an upper Hessenberg matrix [4] that has zeros in the lower triangular part, apart from the very first subdiagonal. Introducing

$$\underline{x}_i = V_i \underline{y}_i$$

the residual norm to be minimized is

$$||A\underline{x}_i - \underline{b}|| = ||AV_i\underline{y}_i - \underline{b}|| = ||V_iH_i\underline{y}_i - V_i(||\underline{r}_0||)\underline{e}_1||.$$

In the latter part of this equation we relied on the Ritz–Galerkin technique introduced in Section 9.3. Since the V_i are orthonormal, the norm simplifies to

$$||H_i\underline{y}_i - (||\underline{r}_0||)\underline{e}_1||.$$

This is now a least squares problem with an upper Hessenberg matrix, for which the technique discussed in Section 9.7 may be efficiently applied. Accordingly, we factorize

$$H_i = Q_i R_i.$$

Then,

$$||H_i\underline{y}_i - (||\underline{r}_0||)\underline{e}_1|| = ||Q_iR_i\underline{y}_i - (||\underline{r}_0||)\underline{e}_1||.$$

Now exploiting the orthonormality of Q_i we obtain

$$||R_i\underline{y}_i - Q_i^T(||\underline{r}_0||)\underline{e}_1||.$$

The norm is minimal when the quantity inside is zero; hence the minimum residual solution is

$$y_i = R_i^{-1}Q_i^T||\underline{r}_0||\underline{e}_1.$$

The inverse is easily executed by backward substitution, since R_i is upper triangular. The next iterative solution is

$$\underline{x}_i = V_i\underline{y}_i.$$

A practical implementation of this class is the GMRES method [14]. While it is a subject of significant academic interest, it is less frequently used in industry.

9.9   Algebraic multigrid method

The final technique described in this chapter has only about 25 years of history and it is still the subject of intense research. The technique of multigrid, as the name implies, has some foundation in the finite difference technique to be discussed in the last chapter. It is based on a two-level discretization of a boundary value problem, one of a fine and another of a coarse grid.

The process starts by obtaining an initial approximate solution on the fine discretization level $\underline{x}_f$ and proceeds with the following steps.

1. Compute the residual of the current approximate solution on the fine grid

$$\underline{r}_f = \underline{b} - A_f \underline{x}_f.$$

2. Transfer the residual to the coarse grid

$$\underline{r}_c = I_c^f \underline{r}_f.$$

3. Solve the coarse grid problem

$$A_c \underline{y}_c = \underline{r}_c.$$

4. Transfer the coarse grid solution to the fine grid

$$\underline{y}_f = I_f^c \underline{y}_c.$$

5. Correct the fine grid solution

$$\underline{x}_f = \underline{x}_f + \underline{y}_f.$$

Several comments are in order. The fine to coarse translation is executed as

$$A_c = I_c^f A_f I_f^c,$$

and conversely, the coarse to fine translation is

$$A_f = I_f^c A_c I_c^f.$$

The I_f^c is the coarsening and the I_c^f the smoothing matrix. They are sometimes called the restriction and prolongation matrices, respectively. The process is repeated until, following step 1, the residual norm dips below an acceptance threshold.

$$||\underline{r}_f|| \le \epsilon.$$

The coarse grid problem posed in step 3 may also recursively be solved by ever coarser application of the same algorithm, thus the name of multigrid as opposed to bigrid.

Finally, for the case of a straightforward linear algebraic problem with a matrix A where the underlying physical problem is not available, the algebraic variation of the multigrid method (AMG) is often used. The naming convention refers to the fact that the coarse subset of the matrix is computed based on algebraic concepts.

An unknown component is a good candidate for being in the coarse subset if it is strongly coupled to another solution component. In that sense, if the $A(i, j)$ component of matrix A is large in comparison to the other off-diagonal terms of the ith row, then $\underline{x}(j)$ is strongly coupled to $\underline{x}(i)$. Hence the jth row will be retained in the coarse set if

$$|A(i, j)| > max_{k \neq i}|A(i, k)|$$

for $i = 1, \ldots, n$. In other words, for each row i of the matrix we gather the j column indices into the coarse set corresponding to the dominant terms in absolute magnitude. This is rather heuristic and some improvements are possible by incorporating weights or using linear combinations between certain components.

The coarsening and the smoothing transformation matrices contain zeroes and ones and some weights, hence their notation with I, and are related as

$$(I_f^c)^T = I_c^f.$$

They have f rows and c columns and vice versa. For example, a coarsening matrix with linear interpolation for a fine matrix of order 5 may be of the form

$$I_f^c = I_5^3 = \begin{bmatrix} c_1 & 0 & 0 \\ 1 & 0 & 0 \\ c_3 & c_3 & 0 \\ 0 & 1 & 0 \\ 0 & c_5 & c_5 \end{bmatrix},$$

where the c_i are interpolation coefficients. A smoothing matrix with some weights w_i may be written as

$$I_c^f = I_3^7 = \begin{bmatrix} w_1 & 1 & w_3 & 0 & 0 \\ 0 & 0 & w_3 & 1 & w_5 \\ 0 & 0 & 0 & 0 & w_5 \end{bmatrix}.$$

For more on the topic, see [13].

9.10   Linear static analysis application

The most frequent practical application of the approximate solution of linear systems in engineering practice is the linear static analysis of a mechanical system. The problem is of the form

$$KU = F,$$

where now the load is a matrix as the structures are usually analyzed under a multitude of load conditions. Hence,

$$F = \begin{bmatrix} \underline{f}_1 \ \underline{f}_2 \ \cdots \ \underline{f}_m \end{bmatrix}$$

and consequently,

$$U = \begin{bmatrix} \underline{u}_1 \ \underline{u}_2 \ \cdots \ \underline{u}_m \end{bmatrix}.$$

The number of load conditions m is in the hundreds and sometimes exceeds a thousand. The size n of the stiffness matrix K is dependent upon the type of the geometric model and the level of discretization, but more often than not is in the millions.

Due to the finite element origin of these problems, it is a natural idea to take advantage of information from the finite element basis.

$$P = f(k_i, N_j), i = 1, \ldots, n_{elements}, j = 1, \ldots, n_{nodes},$$

where k_i are the individual element matrices and N_j are the shape functions. Such preconditioners are called element based preconditioners [5].

While the iterative solutions may be executed simultaneously for multiple load vectors, their cost effectiveness diminishes with very large numbers of right-hand sides. Still, in the case of a very large n, solving the problem by computing direct factorizations may just be impossible due to resource limitations of the computer.

The most attractive case for the use of iterative solutions in linear static analyses is when the finite element model is highly connected, resulting in rather dense matrices. Such models are generated by automatic meshing techniques applied to compact structures, like automobile engine components.

References

[1] Gauss, C. F.; *Werke*, Teubner, Leipzig, 1903.

[2] Golub, G. H. and Van Loan, C. F.; *Matrix Computations*, Johns Hopkins Press, Baltimore, 1983.

[3] Hestenes, M. R. and Stiefel, E.; Methods of conjugate gradients for solving linear systems, *Journal Res. National Bureau of Standards*, Vol. 49, pp. 409–436, 1952.

[4] Householder, A. S.; *The Theory of Matrices in Numerical Analysis*, Blaisdell, New York, 1964.

[5] Hughes, T. J. R., Levit, I. and Winget, J.; An clement by element solution algorithm for problems of structural and solid mechanics, *J. Comp. Methods in Appl. Mech. Eng.*, Vol. 36., pp. 241–154, 1983.

[6] Jacobi, C. G. J.; Über eines leichtes Verfahren die in der Theorie der Sécularstörungen vorkommenden Gleichungen numerisch aufzulösen, *J. Reine Angewandte Math.*, Vol. 30., pp. 51–94, 1846.

[7] Kahan, W.; *Gauss–Seidel Methods of Solving Large Systems of Linear Equations*, University of Toronto, 1958.

[8] Kolotilina, L. Y. and Yeremin, A. Y.; Factorized sparse approximate inverse preconditionings, *J. of Matrix Analysis and Applications*, Vol. 14, pp. 45–58, 1988.

[9] Krylov, A. N.; On the numerical solution of equations which are determined by the frequency of small vibrations of material systems, *Izv. Akad. Nauka*, Vol. 1, pp. 491–539, 1931.

[10] Ritz, W.; Über eine neue Method zur Lösung Gewisser Variationsprobleme der Mathematischen Physik, *J. Reine Angewandte Math.*, Vol. 135, pp. 1–61, 1909.

[11] Parlett, B. N. and Taylor, D.; A look-ahead Lanczos algorithm for unsymmetric matrices, *Mathematics Department Reports*, University of California at Berkeley, 1983.

[12] Rózsa, P.; *Lineáris algebra és alkalmazásai*, Müszaki Kiadó, Budapest, 1974.

[13] Rüde, U.; Fully adaptive multigrid methods, *SIAM Journal of Num. Anal.*, Vol. 30, pp. 230–248, 1993.

[14] Saad, Y. and Schultz, M. H.; GMRES: a generalized minimal residual algorithm for solving nonsymmetric linear systems, *SIAM J. of Sci. Stat. Comp.*, Vol. 7, pp. 856–869, 1986.

[15] Varga, R. S.; *Matrix Iterative Analysis*, Prentice-Hall, Englewod Cliffs, NJ, 1962.

10

Approximate solution of eigenvalue problems

The eigenvalue problem has a hundred-year-old historical foundation dating from Rayleigh [12]. The most profound influence in the area, however, came from Lanczos' heralded paper [9] in 1950. Key ideas of his approach were used in the conjugate gradient method of the last chapter. Anecdotal evidence even suggests that Hestenes and Stiefel gained significant benefit from the presence of Lanczos, who was their coworker at the National Bureau of Standards. The other important development in eigenvalue solutions, regarded as another classical contribution, was the QR iteration of Francis [4].

10.1 Classical iterations

The first methods used to compute eigenvalues of matrices were introduced by engineers. The very simplest and one of the oldest methods, currently known as the power method, was originally called the Stodola iteration. It is simply based on the observation that the eigenvector corresponding to the dominant (i.e., the largest) eigenvalue of the matrix may be obtained by an iterative sequence

$$\underline{x}_i, i = 1, 2, \ldots .$$

The process starts with a unit vector $\underline{x}_0$ and proceeds as

$$\underline{y}_i = A\underline{x}_{i-1}, i = 1, 2, \ldots .$$

The estimated eigenvalue is chosen to be the maximum element of the current iterative vector

$$\lambda_i = max_{k=1}^n |\underline{y}_i(k)|$$

and the eigenvector is scaled by this value as

$$\underline{x}_i = \frac{\underline{y}_i}{\lambda_i}.$$

Note this is not the ith eigenvalue, it is the ith iterate of the dominant eigenvalue λ_{max}. We denote the largest eigenvalue by λ_{max} and the second largest by λ_{max-1}. The process converges with a rate of

$$|\lambda_{max} - \lambda_i| = O((\frac{\lambda_{max-1}}{\lambda_{max}})^i).$$

The above equation implies that if the dominant eigenvalue is well separated, the process is acceptable. Note that the number of iterations i is bounded by ∞, meaning that more than n iterations may be needed to find the dominant eigenvalue. The method in this case, of course, loses any practicality.

In engineering applications the smallest eigenvalue is of more practical interest. This fact gave birth to another classical method, the inverse iteration, a conceptually similar process. We can, however, avoid the inversion of the matrix: instead of multiplying in each step, we solve

$$A\underline{y}_i = \underline{x}_{i-1}, i = 1, 2, \ldots.$$

This process will converge to the eigenvalue closest to the origin, with convergence related to the ratio

$$|\lambda_i - \lambda_{min}| = O((\frac{\lambda_{min}}{\lambda_{min+1}})^i).$$

Again, the separation of the smallest eigenvalue λ_{min} from the second smallest λ_{min+1} defines the rate of convergence.

10.1.1 The technique of deflation

The shortcoming of both of the above classical methods is that they find a single eigenvalue, the largest or the smallest, in one application of the process. The question arises: how to find the next eigenvalue or even more?

In theory, we can set the starting vector of a subsequent iteration, such that it does not have a component in the direction of the already found x_{min} or x_{max} eigenvectors. This approach, however, does not guarantee that during the iteration process, due to the round off and computational errors, the original eigenvector components will not reappear. This would result in the same solution again.

The remedy is to reduce (deflate) the rank of the matrix by a one dimensional subspace representing the already known solution. The following, also classical, method is due to Hotelling [7]. We first consider a symmetric matrix for simplicity of the discussion and assume that we obtained an eigenvalue λ_1

whose eigenvector is normalized so that

$$x_1^T x_1 = 1.$$

We compute a new matrix

$$A_1 = A - \lambda_1 x_1 x_1^T.$$

This matrix has very specific characteristics since

$$(A - \lambda_1 x_1 x_1^T) x_j = A x_j - \lambda_1 x_1 x_1^T x_j = \lambda_j x_j - \lambda_1 x_1 (x_1^T x_j).$$

For the case of $j = 1$ this yields

$$(A - \lambda_1 x_1 x_1^T) x_1 = \lambda_1 x_1 - \lambda_1 x_1 (x_1^T x_1) = 0 x_j.$$

For the case of $j \neq 1$ this yields

$$(A - \lambda_1 x_1 x_1^T) x_j = \lambda_j x_j - \lambda_1 x_1 (0) = \lambda_j x_j = A_1 x_j.$$

The interpretation of these results is that the eigenvalues and corresponding eigenvectors of A_1 are also those of the A matrix. The eigenvalue used in the deflation becomes zero in the new matrix.

The method is also applicable to nonsymmetric matrices in the form of

$$A_1 = A - \lambda_1 x_1 y_1^T$$

where y denotes the left eigenvectors that are orthonormalized as

$$y_1^T x_1 = 1.$$

It is also possible to remove more than one rank one term simultaneously as

$$A_{12} = A - \lambda_1 x_1 y_1^T - \lambda_2 x_2 y_2^T.$$

The method is, however, not very efficient for very large problems and has numerical issues when the eigenvalues are not simple. It was mainly presented here for the conceptual understanding of handling this problem. The more advanced methods discussed later in this chapter have embedded deflation techniques.

10.1.2 Computational example

We consider the unsymmetric matrix

$$A = \begin{bmatrix} 1 & 1 & 2 \\ -1 & 3 & 2 \\ -1 & 2 & 3 \end{bmatrix}$$

whose eigenvalues are $\lambda_1 = 1, \lambda_2 = 2$ and $\lambda_3 = 4$. The corresponding orthonormalized right and left eigenvector pairs are

$$x_1 = \frac{1}{\sqrt{3}} \begin{bmatrix} -2 \\ -2 \\ 1 \end{bmatrix}, y_1 = \frac{1}{\sqrt{3}} \begin{bmatrix} 0 \\ -1 \\ 1 \end{bmatrix},$$

$$x_2 = \frac{1}{\sqrt{2}} \begin{bmatrix} 3 \\ 1 \\ 1 \end{bmatrix}, y_2 = \frac{1}{\sqrt{2}} \begin{bmatrix} 1 \\ -1 \\ 0 \end{bmatrix},$$

and

$$x_3 = \frac{1}{\sqrt{6}} \begin{bmatrix} 1 \\ 1 \\ 1 \end{bmatrix}, y_3 = \frac{1}{\sqrt{6}} \begin{bmatrix} -3 \\ 5 \\ 4 \end{bmatrix}.$$

Executing the deflation by the first eigenvalue we obtain

$$A_1 = \begin{bmatrix} 1 & 1 & 2 \\ -1 & 3 & 2 \\ -1 & 2 & 3 \end{bmatrix} - 1 \cdot \frac{1}{3} \begin{bmatrix} -2 \\ -2 \\ 1 \end{bmatrix} \begin{bmatrix} 0 & -1 & 1 \end{bmatrix} = \begin{bmatrix} 1 & 1/3 & 8/3 \\ -1 & 7/3 & 8/3 \\ -1 & 7/3 & 8/3 \end{bmatrix}.$$

The eigenvalues of this matrix are $\lambda_1 = 0, \lambda_2 = 2$ and $\lambda_3 = 4$, which shows that the deflated eigenvalue became zero, but the others remained. The same iteration method then may be used on this matrix to find the next eigenvalue.

Deflating the original matrix with $\lambda_2 = 2$ will result in

$$A_2 = \begin{bmatrix} -2 & 4 & 2 \\ -2 & 4 & 2 \\ -2 & 3 & 3 \end{bmatrix}.$$

Note that the matrix still has three eigenvalues, $0, 1, 4$, from which the zero represents the current deflation by $\lambda_2 = 2$ and the others are the eigenvalues of the original matrix. The remaining third eigenvalue $\lambda_3 = 4$ may be found from this matrix, but it is simpler to use the simultaneous deflation by the first two eigenvalues. This results in the matrix

$$A_{12} = \begin{bmatrix} -2 & 10/3 & 8/3 \\ -2 & 10/3 & 8/3 \\ -2 & 10/3 & 8/3 \end{bmatrix}$$

whose eigenvalues are $0, 0, 4$, the two zeroes representing the simultaneously deflated eigenvalues and the last eigenvalue is now easier to find.

It is noticeable that the A_1, A_2 matrices both have one rank less than the original since two rows are linearly dependent as they are identical. The A_{12} matrix is then of rank one because all rows are identical, hence linearly dependent, due to the fact that two deflations were executed simultaneously.

10.1.3 Improving the eigenvalue accuracy

Let us now assume that we approximately (indicated by tilde) solved the eigenvalue problem of an unsymmetric matrix for which both right- and left-hand eigenvectors exist:

$$A\tilde{u}_i = \tilde{\lambda}_i \tilde{u}_i$$

and

$$A^T \tilde{v}_i = \tilde{\lambda}_i \tilde{v}_i.$$

We write the (assumedly small) errors of the approximation as

$$\lambda_i = \tilde{\lambda}_i + \Delta\lambda_i,$$
$$u_i = \tilde{u}_i + \Delta u_i$$

and

$$v_i^T = \tilde{v}_i^T + \Delta v_i^T.$$

The exact right-hand side solution would be

$$Au_i = \lambda_i u_i.$$

Substituting results in

$$A\tilde{u}_i + A\Delta u_i - \tilde{\lambda}_i \tilde{u}_i + \tilde{\lambda}_i \Delta u_i + \Delta\lambda_i \tilde{u}_i + \Delta\lambda_i \Delta u_i.$$

Following [2], omitting the second order small last term and introducing the residual of the right-hand approximate solution

$$r_i = A\tilde{u}_i - \tilde{\lambda}_i \tilde{u}_i,$$

we obtain

$$A\Delta u_i - \tilde{\lambda}_i \Delta u_i \approx -r_i + \Delta\lambda_i \tilde{u}_i.$$

Premultiplying with a left-hand solution results in

$$\tilde{v}_j^T A\Delta u_i - \tilde{\lambda}_i \tilde{v}_j^T \Delta u_i \approx -\tilde{v}_j^T r_i + \Delta\lambda_i \tilde{v}_j^T \tilde{u}_i.$$

In the case of a reasonably accurate approximation

$$\tilde{v}_j^T A \approx \tilde{\lambda}_j \tilde{v}_j^T,$$

hence it follows that

$$(\tilde{\lambda}_j - \tilde{\lambda}_i)\tilde{v}_j^T \Delta u_i \approx -\tilde{v}_j^T r_i + \Delta\lambda_i \tilde{v}_j^T \tilde{u}_i.$$

Assuming $i = j$ and that the eigenvalues are all simple we obtain a correction formula for the eigenvalue as

$$\Delta\lambda_i \approx \frac{\tilde{v}_i^T r_i}{\tilde{v}_i^T \tilde{u}_i}.$$

To obtain a correction formula for the right eigenvectors, we write the error vector as

$$\Delta u_i = \sum_{j \neq i, j=1}^{n} h_{ij} \tilde{u}_j,$$

where the h_{ij} constants are small numbers. Considering the above equation now for $i \neq j$, exploiting that $\tilde{v}_j^T \Delta u_i \approx h_{ij} \tilde{v}_j^T \tilde{u}_i$ and $\Delta \lambda_i \tilde{v}_j^T \tilde{u}_i \approx 0$, we obtain

$$h_{ij} = \frac{\tilde{v}_j^T r_i}{(\tilde{\lambda}_i - \tilde{\lambda}_j) \tilde{v}_j^T \tilde{u}_j}.$$

Hence the corrected (better, but not necessarily exact) right eigenvector becomes

$$u_i = \tilde{u}_i + \sum_{j \neq i, j=1}^{n} h_{ij} \tilde{u}_j.$$

Similarly by introducing the left-hand residual of

$$\rho_i^T = \tilde{v}_i^T A - \tilde{\lambda}_i \tilde{v}_i^T,$$

and the left-hand error vector as

$$\Delta v_i^T = \sum_{j \neq i, j=1}^{n} g_{ij} \tilde{v}_j^T,$$

the coefficients become

$$g_{ij} = \frac{\rho_i^T \tilde{u}_j}{(\tilde{\lambda}_i - \tilde{\lambda}_j) \tilde{v}_j^T \tilde{u}_j}.$$

These computations now tend to be rather tedious, hence most of the time only the eigenvalue correction formula is used.

10.1.4   Computational example

Let us illustrate the method of the last section with the unsymmetric matrix

$$A = \begin{bmatrix} 2 & 3 & 2 \\ 10 & 3 & 4 \\ 3 & 6 & 1 \end{bmatrix}$$

whose maximum eigenvalue is $\lambda_1 = 11$. The corresponding left and right eigenvectors are

$$u_1 = \begin{bmatrix} 2/3 \\ 4/3 \\ 1 \end{bmatrix}$$

and

$$v_1 = \begin{bmatrix} 4/3 \\ 1 \\ 2/3 \end{bmatrix},$$

since

$$\frac{v_1^T (Au_1)}{v_1^T u_1} = 11.$$

For simplicity we create hypothetical approximate vectors by adding 10^{-1} to each term of the exact vectors:

$$\tilde{u}_1 = \begin{bmatrix} 2/3 + 1/10 \\ 4/3 + 1/10 \\ 1 + 1/10 \end{bmatrix}$$

and

$$\tilde{v}_1 = \begin{bmatrix} 4/3 + 1/10 \\ 1 + 1/10 \\ 2/3 + 1/10 \end{bmatrix}.$$

Let us assume that the approximate eigenvalue is also perturbed similarly but by a smaller value as

$$\tilde{\lambda}_1 = 11 + 1/100 = 11.01.$$

The residual of this approximate solution is

$$r_1 = \begin{bmatrix} -0.40767 \\ 0.58567 \\ -0.1110 \end{bmatrix}.$$

Evaluating the eigenvalue correction term we obtain

$$\Delta\lambda_1 = -0.007158.$$

Adding this to the approximated eigenvalue,

$$\lambda_1 = 11.01 - 0.007 = 11.003,$$

which is clearly much closer to the exact eigenvalue, the error decreased by an order of magnitude. Applying now the eigenvector correction forms also shown above, but not computed here, new approximate eigenvectors are found. Using them and repeating the above process will correct the eigenvalue even further.

While the processes in the last sections will ultimately converge to the correct solution, in the following sections we will discuss methods in which both the deflation and corrections are already built in, hence producing better solutions more efficiently.

10.2   The Rayleigh–Ritz procedure

As foundation, we start from another classical method, the so-called Rayleigh quotient iteration. It is based on the late 19th century effort by Lord Rayleigh [12], who approximated the first mode of a vibrating system by solving

$$(A - \rho(\underline{x}_1))\underline{x}_1 = \underline{e}_1.$$

The term ρ is now called the Rayleigh quotient. For a matrix A and vector $\underline{x}$, it is the scalar

$$\rho(\underline{x}) = \frac{\underline{x}^T A \underline{x}}{\underline{x}^T \underline{x}}.$$

It is easy to deduce that if $\underline{x}$ is an eigenvector of A, then the Rayleigh quotient is the corresponding eigenvalue

$$\lambda = \rho(\underline{x}).$$

This observation leads to the approximation technique known as the Rayleigh quotient iteration. The process starts as before and in each step of $i = 1, 2, \ldots$ we compute an approximate eigenvalue from the Rayleigh quotient

$$\lambda_i = \frac{\underline{x}_{i-1}^T A \underline{x}_{i-1}}{\underline{x}_{i-1}^T \underline{x}_{i-1}}.$$

Then we solve for an approximate eigenvector as

$$(A - \lambda_i I)\underline{y}_i = \underline{x}_{i-1}.$$

We normalize the approximate eigenvector as

$$\underline{x}_i = \frac{\underline{y}_i}{||\underline{y}_i||},$$

and the process continues. The process converges globally with a cubic rate of convergence, as shown in [10].

Let us now restrict our focus to the linear symmetric eigenvalue problem of order n

$$A\underline{x}_j = \lambda_j \underline{x}_j.$$

Here $j = 1, 2, \ldots, n$, although in practice the number of eigenvalues computed is usually much less. Let us assume that we have a matrix Q with m orthonormal columns

$$Q^T Q = I.$$

The Rayleigh–Ritz procedure computes the matrix Rayleigh quotient of

$$T_m = Q^T A Q.$$

This matrix has m rows and columns and is tridiagonal if A is symmetric and Q spans a Krylov subspace. The eigenvalues of this matrix are called the Ritz values and are computed from the Ritz problem of

$$T_m \underline{u}_j = \theta_j \underline{u}_j.$$

The approximate eigenvectors (also called Ritz vectors) of the original eigenvalue problem are recovered as

$$\underline{x}_j = Q\underline{u}_j.$$

This procedure, in essence, provides an approximate solution to the eigenvalue problem by projecting the matrix into a subspace spanned by the columns of Q. The residual error of the approximate eigenvector is computed as

$$\underline{r}_j = r(\underline{x}_j) = A\underline{x}_j - \theta_j \underline{x}_j = AQ\underline{u}_j - \theta_j \underline{x}_j.$$

This error bounds the eigenvalue as

$$(\theta_j - ||r_j||) \leq \lambda_j \leq (\theta_j + ||r_j||).$$

The quality of the Rayleigh–Ritz procedure depends on the selection of the subspace spanned by Q. The performance of the method depends on the construction of the subspace. The aforementioned Krylov subspace is a good selection; however, computing it efficiently is where the Lanczos method exceptionally shines.

10.3   The Lanczos method

The Lanczos method [9] is an excellent way of generating the Krylov subspace of A and projecting the eigenvalue problem accordingly. The problem we consider is

$$A\underline{x} = \lambda \underline{x}$$

where A is real, symmetric and $\underline{x}$ are the eigenvectors of the original problem; the underlining is used to distinguish from the soon to be introduced x Lanczos vectors which, for the sake of clarity, will not be underlined.

This method generates a set of n orthonormal vectors, the Lanczos vectors X_n, such that

$$X_n^T X_n = I,$$

where I is the identity matrix of order n and

$$X_n^T A X_n = T_n.$$

Here we ignore the possibility of a numerical breakdown of this process. The T_n order n tridiagonal matrix is of the form

$$T_n = \begin{bmatrix} \alpha_1 & \beta_1 & & & \\ \beta_1 & \alpha_2 & \beta_2 & & \\ & \cdot & \cdot & \cdot & \\ & & \beta_{n-2} & \alpha_{n-1} & \beta_{n-1} \\ & & & \beta_{n-1} & \alpha_n \end{bmatrix}.$$

The terms of the tridiagonal matrix will be computed as the coefficients of the Lanczos recursion process.

By appropriate premultiplication, we get

$$AX_n = X_n T_n.$$

By equating columns in this equation, we get

$$Ax_k = \beta_{k-1} x_{k-1} + \alpha_k x_k + \beta_k x_{k+1},$$

where $k = 1, 2, ..n - 1$ and x_k are the kth columns of X_n. For any $k < n$ the following matrix form describes the recurrence.

$$AX_k = X_k T_k + \beta_k x_{k+1} e_k^T,$$

where e_k is the kth unit vector containing one in row k and zeroes elsewhere. Its presence is only needed to make the matrix addition operation compatible.

By reordering, we obtain the following famous Lanczos recurrence formula:

$$\beta_k x_{k+1} = A x_k - \alpha_k x_k - \beta_{k-1} x_{k-1}.$$

The coefficients β_k and α_k are computed as follows. The successive Lanczos vectors are orthogonal:

$$x_k^T x_j = 0; j = 1, 2, \ldots, k - 1,$$

and they are also normalized as

$$||x_k|| = 1.$$

This is the source of the β_k coefficients. Then we premultiply the earlier equation by x_k^T.

$$x_k^T A x_k = \beta_{k-1} x_k^T x_{k-1} + \alpha_k x_k^T x_k + \beta_k x_k^T x_{k+1}.$$

Based on the orthonormality of the Lanczos vectors, this yields

$$\alpha_k = x_k^T A x_k,$$

which is the second coefficient. Now, we need to find the eigenvalues and eigenvectors of this tridiagonal matrix.

$$T_n u_j = \theta_j u_j.$$

The j in the above equation is the index of the eigenvalue of the tridiagonal matrix, $j = 1, 2, \ldots n$.

Since the eigenvalues of A are invariant under the transformation to tridiagonal form, the θ eigenvalues of the tridiagonal matrix T_n are the same as the λ eigenvalues of the original matrix. The eigenvectors of the original problem are calculated from the eigenvectors of the tridiagonal problem by the procedure originally suggested by Lanczos:

$$\underline{x}_j = X_n u_j,$$

where X_n is the matrix containing the n Lanczos vectors, u_j is the jth eigenvector of the tridiagonal matrix.

10.3.1 Truncated Lanczos process accuracy

The Lanczos iteration may also be stopped at step j, resulting in a tridiagonal matrix T_j. This is especially important when dealing with very large matrix sizes commonly occurring in engineering practice. The method still enables the computation of approximate eigenvalues and eigenvectors as follows.

The θ eigenvalues of the tridiagonal matrix T_j, the Ritz values, are approximations of the λ eigenvalues of the original matrix. The eigenvectors of the original problem are calculated from the approximate eigenvectors (Ritz vectors) as above.

The approximated residual error in the original solution [10] is

$$||\underline{r}_j|| = ||A\underline{x}_j - \lambda\underline{x}_j|| = ||AX_iu_j - \lambda_jX_iu_j|| = ||(AX_i - \lambda_jX_i)u_j||,$$

where $j = 1, 2, \ldots i$, and i is the number of Lanczos steps executed. It follows that

$$||\underline{r}_j|| = ||(AX_i - X_iT_i)u_j|| = ||(\beta_ix_{i+1}e_i^T)u_j|| = \beta_i||e_i^T||\,||u_j||,$$

since the norm of the Lanczos vector x_{i+1} is unity. Taking advantage of the structure of the unit vector, we can simplify into the following scalar form:

$$||\underline{r}_j|| = \beta_i|u_{ij}|,$$

where u_{ij} is the ith (last) term in the u_j eigenvector. The last equation gives an approximation monitoring tool. When the error norm is less than the required tolerance ϵ, then the jth approximate eigenpair may be accepted.

It is the beauty of this convergence criterion that only the eigenvector of the tridiagonal problem has to be found, which is inexpensive compared to finding the eigenvector of the original (size n) problem.

10.4   The solution of the tridiagonal eigenvalue problem

The eigenvalues of the tridiagonal problem may be bounded and counted by the Sturm sequence property, introduced in Section 7.3.1, and applied to our case as follows. Define polynomials $p_j(x)$ such that

$$p_j(x) = det(T_j - Ix),$$

where T_j is the jth principal minor of the tridiagonal matrix as

$$T_j = \begin{bmatrix} t_{1,1} & t_{1,2} & & \\ t_{2,1} & t_{2,2} & t_{2,3} & \\ & \cdot & \cdot & \cdot \\ & & t_{j,j-1} & t_{j,j} \end{bmatrix}.$$

Starting from $p_0(x) = 1$, the recursion

$$p_j(x) = (t_{j,j} - x)p_{j-1}(x) - t_{j,j-1}^2 p_{j-2}(x),$$

defines this sequence. The number of eigenvalues that are smaller than a certain value $x = \lambda_s$ is the same as the number of sign changes in the sequence

$$p_0(\lambda_s), p_1(\lambda_s), p_2(\lambda_s), \ldots, p_n(\lambda_s).$$

To find an approximate solution to the tridiagonal eigenvalue problem, Francis's QR iteration [4] may be used. The QR method is based on a decomposition of the T_n matrix into the form

$$T_n = Q^1 R^1.$$

R^1 is an upper triangular matrix and Q^1 contains orthogonal columns.

$$Q^{1,T} Q^1 = I.$$

Hence,

$$Q^{1,T} T_n = R^1,$$

and postmultiplying yields

$$Q^{1,T} T_n Q^1 = R^1 Q^1.$$

Repeated execution on the left-hand side produces the Rayleigh quotient (with a unit denominator due to the orthogonality of Q^i)

$$Q^{i,T} Q^{i-1,T} \ldots Q^{2,T} Q^{1,T} T_n Q^1 Q^2 \ldots Q^{i-1} Q^i = \Lambda^i,$$

yielding the eigenvalues.

To accelerate the convergence, a shifted version of this process is executed. One accepted and practical shift is chosen by using the last term of the tridiagonal matrix

$$\omega = T_n(n, n).$$

Then the process is executed on the shifted matrix

$$T_n - \omega I = Q^1 R^1,$$

where ω is the shift. The shifted QR iteration proceeds as

$$T_n^1 = R^1 Q^1 + \omega I.$$

The newly created matrix preserves the eigenvalue spectrum of the old one. By repeatedly applying this procedure, T_n^i converges to diagonal form, with elements that are the approximate eigenvalues of the original matrix.

$$T_n^i = \Lambda^i.$$

The number of iterations i is usually $O(n)$. The computation of Francis' QR iteration takes advantage of the tridiagonal nature of T_n and preserves it in the following iterates.

During the execution of the process, the subdiagonal terms are constantly monitored, and in case any of them becomes sufficiently small, the problem is decoupled into two parts. If one of the partitions is of size one or two, the related eigenvalue problem is solved analytically. The example in the next section will demonstrate this process.

For eigenvectors, an inverse power iteration procedure originally proposed by Wilkinson [15] is commonly used. The eigenvectors corresponding to the jth eigenvalue of the T_n tridiagonal matrix may be determined by the factorization

$$T_n - \theta_j I = L_j U_j,$$

where L_j is unit lower triangular (the diagonal terms are one) and U_j is upper triangular. The θ_j term is an approximation of the jth eigenvalue. Gaussian elimination with partial pivoting is used, i.e., the pivotal row at each stage is selected to be the equation with the largest coefficient of the variable being eliminated. Since the original matrix is tridiagonal, at each stage there are only two equations containing that variable. Approximate eigenvectors of the jth eigenvalue θ_j will be calculated by a simple (since the U_j also has only 2 codiagonals) iterative procedure.

$$U_j \underline{u}_j^{i+1} = \underline{u}_j^i,$$

where $\underline{u}_j^i$ is random and i is the iteration counter. Practice shows that the convergence of this procedure is so rapid that i only goes to 2 or 3.

This original method may also be extended to deal with special cases such as repeated eigenvalues. There are also more recent approaches, such as described in [3], that combine the two steps together; however, their industrial value has not been fully established yet.

10.5   The biorthogonal Lanczos method

Lanczos himself originally described his method for unsymmetric matrices. In this case a biorthogonal version of the method is executed with a pair of Lanczos vectors at each iteration step.

$$\beta_k x_{k+1} = A x_k - \alpha_k x_k - \gamma_{k-1} x_{k-1},$$

and

$$\gamma_k y_{k+1}^T = y_k^T A - \alpha_k y_k^T - \beta_{k-1} y_{k-1}^T,$$

which results in an unsymmetric tridiagonal matrix built from the Lanczos coefficients as follows:

$$T_n = \begin{bmatrix} \alpha_1 & \gamma_1 & & & \\ \beta_1 & \alpha_2 & \gamma_2 & & \\ & & \cdot & \cdot & \cdot & \\ & & & \beta_{n-2} & \alpha_{n-1} & \gamma_{n-1} \\ & & & & \beta_{n-1} & \alpha_n \end{bmatrix}.$$

The coefficients β_k and γ_k may be defined such that

$$\beta_k \gamma_k = \overline{y}_{k+1}^T \overline{x}_{k+1},$$

is satisfied, where

$$\overline{x}_{k+1} = \beta_k x_{k+1},$$

and

$$\overline{y}_{k+1}^T = \gamma_k y_{k+1}^T.$$

Introducing the normalization parameter

$$\delta_k = \overline{y}_{k+1}^T \overline{x}_{k+1},$$

the coefficients are selected as

$$\beta_k = \sqrt{|\delta_k|},$$

and

$$\gamma_k = \beta_k \mathrm{sign}(\delta_k).$$

The advantage of these choices is that the resulting tridiagonal matrix will be symmetric, except for the sign. In exact arithmetic every pair of Lanczos vectors satisfies

$$y_k^T x_k = 1,$$

and

$$y_k^T x_j = 0, \quad y_j^T x_k = 0,$$

for any $j < k$. Based on this orthonormality condition, premultiplying results in

$$\alpha_k = y_k^T A x_k,$$

which is, of course, the equation to produce the diagonal terms of the tridiagonal matrix.

10.5.1   Computational example

Let us find the eigenvalues and eigenvectors of the following order 3 unsymmetric matrix.

$$A = \begin{bmatrix} 1/2 & 1/2 & -1/2 \\ 0 & 0 & -2 \\ 3/2 & -1/2 & 9/2 \end{bmatrix}.$$

We use the starting vectors:

$$x_1 = \begin{bmatrix} 1 \\ 0 \\ -1 \end{bmatrix},$$

and

$$y_1 = \begin{bmatrix} 1/2 \\ -1/2 \\ -1/2 \end{bmatrix},$$

which satisfy

$$y_1^T x_1 = 1.$$

Step $k = 1$:

We first compute the diagonal Lanczos coefficient as

$$\alpha_1 = y_1^T A x_1 = 1.$$

Since, in the $k = 1$ step, the coefficients with zero indices γ_0, β_0 are zero, the Lanczos recurrence equation is

$$\bar{x}_2 = A x_1 - \alpha_1 x_1 = \begin{bmatrix} 1 \\ 2 \\ -3 \end{bmatrix} - \begin{bmatrix} 1 \\ 0 \\ -1 \end{bmatrix} = \begin{bmatrix} 0 \\ 2 \\ -2 \end{bmatrix}.$$

Similarly for the left-hand side,

$$\bar{y}_2 = A^T y_1 - \alpha_1 y_1 = \begin{bmatrix} -1/2 \\ 1/2 \\ -3/2 \end{bmatrix} - \begin{bmatrix} 1/2 \\ -1/2 \\ 1/2 \end{bmatrix} = \begin{bmatrix} -1 \\ 1 \\ -1 \end{bmatrix}.$$

The normalization parameter is calculated as

$$\delta_1 = \bar{y}_2^T \bar{x}_2 = 4.$$

Since it is positive, its square root becomes the value of the off-diagonal coefficients:

$$\beta_1 = \sqrt{\delta_1} = 2 = \gamma_1.$$

In turn, the normalized Lanczos vectors for the next iteration are

$$x_2 = \bar{x}_2/\beta_1 = \begin{bmatrix} 0 \\ 1 \\ -1 \end{bmatrix},$$

and

$$y_2 = \bar{y}_2/\gamma_1 = \begin{bmatrix} -1/2 \\ 1/2 \\ -1/2 \end{bmatrix}.$$

Step $k = 2$:

The execution of the next step for these new vectors starts again with computing the diagonal coefficient.

$$\alpha_2 = y_2^T A x_2 = 3.$$

The Lanczos recurrence step is now full for this case:

$$\bar{x}_3 = A x_2 - \alpha_2 x_2 - \gamma_1 x_1 = \begin{bmatrix} -1 \\ -1 \\ 0 \end{bmatrix}.$$

Similarly for the left-hand recurrence,

$$\bar{y}_3 = A^T y_2 - \alpha_2 y_2 - \gamma_1 y_1 = \begin{bmatrix} -1/2 \\ -1/2 \\ -1/2 \end{bmatrix}.$$

Since

$$\delta_2 = 1,$$

then

$$\beta_2 = \gamma_2 = 1.$$

The normalized Lanczos vectors of the final iteration are

$$x_3 = \begin{bmatrix} -1 \\ -1 \\ 0 \end{bmatrix},$$

and

$$y_3 = \begin{bmatrix} -1/2 \\ -1/2 \\ -1/2 \end{bmatrix}.$$

Step $k = 3$:

The last diagonal coefficient becomes

$$\alpha_3 = y_3^T A x_3 = 1,$$

and the iteration process stops as we have reached the full size of the matrix. This would be the stage to check the accuracy of the approximate solution in case a Rayleigh–Ritz approximation is executed and the Lanczos process stopped here.

Verification step:

The resulting tridiagonal matrix is produced as indicated by the Lanczos process.

$$T_3 = Y^T A X = \begin{bmatrix} 1/2 & -1/2 & -1/2 \\ -1/2 & 1/2 & -1/2 \\ -1/2 & -1/2 & -1/2 \end{bmatrix} A \begin{bmatrix} 1 & 0 & -1 \\ 0 & 1 & -1 \\ -1 & -1 & 0 \end{bmatrix} = \begin{bmatrix} 1 & 2 & 0 \\ 2 & 3 & 1 \\ 0 & 1 & 1 \end{bmatrix}.$$

The Lanczos vectors also satisfy the $Y^T X = I$ orthogonality criterion as it was part of the computation.

We obtain the eigenvalues by applying the QR iteration for

$$T = \begin{bmatrix} 1 & 2 & 0 \\ 2 & 3 & 1 \\ 0 & 1 & 1 \end{bmatrix}.$$

The shift is chosen as

$$\omega - T(3,3) = 1.$$

The QR factorization of the shifted matrix is

$$T - \omega I = \begin{bmatrix} 0 & 2 & 0 \\ 2 & 2 & 1 \\ 0 & 1 & 0 \end{bmatrix} = \begin{bmatrix} 0 & 2/\sqrt{5} & -1/\sqrt{5} \\ -1 & 0 & 0 \\ 0 & 1/\sqrt{5} & 2/\sqrt{5} \end{bmatrix} \begin{bmatrix} -2 & -2 & -1 \\ 0 & \sqrt{5} & 0 \\ 0 & 0 & 0 \end{bmatrix} = Q^1 R^1.$$

Back-multiplication and reshift results in

$$T^1 = R^1 Q^1 + \omega I = \begin{bmatrix} 3 & -\sqrt{5} & 0 \\ -\sqrt{5} & 1 & 0 \\ 0 & 0 & 1 \end{bmatrix}.$$

There is a decoupling into two parts. The size one part on the lower right corner yields

$$det(1 - \lambda) = 0,$$

or

$$\lambda_1 = 1.$$

The upper left corner, size two part, is solved as

$$det(\begin{bmatrix} 3 - \lambda & -\sqrt{5} \\ -\sqrt{5} & 1 - \lambda \end{bmatrix}) = 0.$$

The solutions are

$$\lambda_2 = 2 + \sqrt{6},$$

and

$$\lambda_3 = 2 - \sqrt{6}.$$

Finally, we apply Wilkinson's eigenvector generation. For λ_1 we use an approximate eigenvalue of $\theta = 0.99999$, thus

$$T - \theta I = \begin{bmatrix} 0.00001 & 2 & 0 \\ 2 & 2.00001 & 1 \\ 0 & 1 & 0.00001 \end{bmatrix}.$$

This matrix is factored with pivoted Gaussian elimination into

$$PLU = \begin{bmatrix} 0 & 1 & 0 \\ 1 & 0 & 0 \\ 0 & 0 & 1 \end{bmatrix} \begin{bmatrix} 1 & 0 & 0 \\ 0 & 1 & 0 \\ 0 & 1/2 & 1 \end{bmatrix} \begin{bmatrix} 2 & 2.00001 & 1 \\ 0 & 1.99999 & 0 \\ 0 & 0 & 0.00001 \end{bmatrix}.$$

Here the matrix P represents the pivoting executed during the elimination. Starting with a unit vector associated with this eigenvalue

$$\underline{u}_1^0 = \begin{bmatrix} 0 \\ 0 \\ 1 \end{bmatrix},$$

a single solution step

$$U\underline{u}_1^1 = \underline{u}_1^0$$

results in

$$u_1^1 = \begin{bmatrix} -4.0 * 10^4 \\ -2.0 * 10^{-1} \\ 8.0 * 10^4 \end{bmatrix}.$$

Normalizing with the first component results in the right eigenvector of the tridiagonal problem as

$$u_1 = \begin{bmatrix} 1 \\ 0 \\ -2 \end{bmatrix}.$$

The corresponding eigenvector of the original problem is computed by

$$\underline{x}_1 = X\underline{u}_1 = \begin{bmatrix} 3 \\ 2 \\ -1 \end{bmatrix}.$$

The left eigenvector corresponding to the unit eigenvalue is obtained similarly.

$$\underline{y}_1 = Y^T u_1 = \begin{bmatrix} 3 \\ 1 \\ 1 \end{bmatrix}.$$

With these, both the

$$A\underline{x}_1 = \lambda_1 \underline{x}_1$$

and

$$A^T \underline{y}_1 = \lambda_1 \underline{y}_1$$

are satisfied. The other two eigenvectors may be computed in the same fashion.

10.6   The Arnoldi method

There is another method applicable to unsymmetric systems, the Arnoldi method [1]. It builds an orthonormal basis for the Krylov subspace $K^i(A, \underline{r}_0)$ by computing

$$\underline{u} = A\underline{v}_i$$

and orthonormalizing this vector with respect to all prior $\underline{v}_i$. The matrix form of this scheme is

$$AV_{i-1} = V_i H_i,$$

where the reduced form achieved is an upper Hessenberg matrix of i rows and $i - 1$ columns.

$$H_i = \begin{bmatrix} h_{1,1} & h_{1,2} & . & h_{1,i-1} \\ h_{2,1} & h_{2,2} & . & h_{2,i-1} \\ 0 & h_{3,2} & . & h_{3,i-1} \\ . & & . & . \\ 0 & 0 & . & h_{i,i-1} \end{bmatrix}.$$

The orthogonalization is executed by the Gram–Schmidt method again, and the terms of the matrix may be computed by the following algorithmic steps.

$\underline{v}_1 = \underline{r}_0 \| \underline{r}_0 \|.$

For $i = 1, 2, \ldots$:

$\underline{u} = A\underline{v}_i.$

For $j = 1, 2, \ldots, i$:

$h_{j,i} = \underline{v}_j^T u,$

$\underline{u} = \underline{u} - h_{j,i}\underline{v}_j.$

End of loop on j.

$h_{i+1,i} = \| \underline{u} \|,$

$\underline{v}_{i+1} = \frac{1}{h_{i+1,i}} \underline{u}.$

End of loop on i.

Based on the complexity of this algorithm and due to the topology of the upper Hessenberg format, the Arnoldi method is inherently less efficient than the tridiagonal Lanczos method. There have been attempts at improving the technique to be competitive with the Lanczos method. Such is the implicitly restarted variant of the method [14]. Despite these efforts, the Lanczos method remains the most widely used in the industry.

10.7   The block Lanczos method

In practical applications, matrices frequently have repeated eigenvalues. This occurs, for example, when there is symmetry in the underlying geometry. The single-vector Lanczos method in exact arithmetic can find only one eigenvector of a set of repeated eigenvalues. An approximate calculation of repeated eigenvalues is possible with a block formulation [6], which executes the Lanczos method with multiple vectors simultaneously.

The block Lanczos method to solve the above eigenvalue problem is formulated as

$$R_{i+1} = AQ_i - Q_iA_i - Q_{i-1}B_i^T \quad i = 1, 2, \ldots j,$$

where

$$A_i = Q_i^T AQ_i,$$

and

$$R_{i+1} = Q_{i+1}B_{i+1}.$$

Q_{i+1} is an n by b matrix with orthonormal columns (the Lanczos vectors) and B_{i+1} is a b by b upper triangular matrix, n being the problem size and b the block size, obtained by the QR decomposition.

At this stage several steps (k) of orthogonalization are required to maintain the process. The orthogonalization against all prior Lanczos vectors, commonly called full orthogonalization, is very expensive and its cost increases as the frequency range of interest widens. One avenue to ease this burden is by executing the orthogonalization only to a certain accuracy, called partial orthogonalization in practice [13].

Another practically used improvement is the selective orthogonalization concept [11], which executes orthogonalization against "selected" eigenvectors that are close enough or for other reasons have a large influence on local convergence. The m vectors to be orthogonalized against are these selectively chosen vectors in the following equation:

$$Q_{i+1}^k = Q_{i+1}^{k-1} - \Sigma_{j=1}^m x_j(x_j^T Q_j^{k-1}).$$

Ultimately the block process results in a block tridiagonal matrix of the form

$$T_j = \begin{bmatrix} A_1 & B_2^T & & \\ B_2 & A_2 & B_3^T & \\ & \cdot & \cdot & \cdot \\ & & B_j & A_j \end{bmatrix}.$$

With appropriately chosen Givens transformations [5] this block tridiagonal matrix is reduced into a scalar tridiagonal matrix T_J. The eigensolution of the size $J = jb$, $(J << n)$ eigenvalue problem of

$$T_J u = \lambda u,$$

may be found by the earlier discussed QR iteration algorithm.

To find the eigenvectors of the original problem a back-transformation of the form

$$\underline{x} = Q_J u$$

is required. The Q_J matrix is a collection of the Lanczos vector blocks:

$$Q_J = \begin{bmatrix} Q_1 \ Q_2 \ .. \ Q_j \end{bmatrix}.$$

10.7.1 Preconditioned block Lanczos method

In line with some of the concepts introduced in linear system solutions, a way to accelerate this technology is by preconditioning the eigenvalue problem. We need a preconditioning matrix P for which

$$PP^T = I.$$

We apply it to the block Lanczos recurrence.

$$P^T Q_{i+1} B_{i+1} - P^T A Q_i - P^T Q_i A_i - P^T Q_{i-1} B_i^T.$$

Introduce a preconditioned Lanczos vector block

$$\overline{Q}_i = P^T Q_i,$$

and a preconditioned matrix

$$\overline{A} = P^T A P.$$

Inserting these yields

$$\overline{Q}_{i+1} B_{i+1} = \overline{A} \, \overline{Q}_i - \overline{Q}_i A_i - \overline{Q}_{i-1} B_i^T.$$

The preconditioned Lanczos recurrence is still correct for the eigenvalue problem since the orthonormality of the preconditioned Lanczos vectors is equivalent to the orthonormality of the Lanczos vectors of the original problem as

$$\overline{Q}_i^T \overline{Q}_i = Q_i^T P P^T Q_i = Q_i^T Q_i = I.$$

The success of this technique lies in the selection of the P matrix. It is dependent upon the geometric modeling technique applied (shell vs. solid models) and other characteristics of the particular engineering problem.

10.8   Normal modes analysis application

The undamped, free vibration of mechanical systems in engineering results in
a generalized linear eigenvalue problem of the form:

$$M\ddot{v}(t) + Kv(t) = 0,$$

where the K and M matrices are the stiffness and mass matrices, respectively.
The v is the displacement vector and $\ddot{v}$ is the acceleration. The motion is time
dependent and usually a complex Fourier transformation is executed to transform it to the frequency domain by

$$u(\omega) = \int_0^\infty v(t)e^{-i\omega t}dt,$$

where ω is the frequency of the vibration. Assuming zero displacement at the
initial time and introducing

$$\lambda = \omega^2$$

results in a so-called generalized (two matrix) linear eigenvalue problem.

$$K\phi - \lambda M\phi = 0.$$

Since the engineer's interest in these problems is at the lower end of the spectrum, it is advisable to execute a step of so-called spectral transformation [8].
The substitution of

$$\mu = \frac{1}{\lambda - \lambda_s}$$

will change the problem into

$$(K - \lambda_s M)\phi = \frac{1}{\mu}M\phi.$$

This may be transformed into

$$\mu\phi = (K - \lambda_s M)^{-1}M\phi = A\phi,$$

which is a canonical form amenable to the Lanczos algorithm discussed earlier.
The matrix in this form is unsymmetric, despite the fact that the finite element
matrices are symmetric. The efficiency improvement of solving a symmetric
problem as opposed to an unsymmetric one is very clear from this chapter. In
the industry, therefore, the form

$$\mu M\phi = M(K - \lambda_s M)^{-1}M\phi$$

is used, which produces mass-orthogonal eigenvectors as

$$\phi^T M \phi = I.$$

This is a desirable characteristic in view of the additional dynamic computations often executed in engineering.

References

[1] Arnoldi, W. E.; The principle of minimized iterations in the solution of matrix eigenvalue problems, *Quarterly of Appl. Math.*, Vol. 9, pp. 17–29, 1951.

[2] Bajcsay, P.; *Numerikus analizis*, Tankönyvkiadó, Budapest, 1972.

[3] Dhillon, I. S. and Parlett, B. N.; Multiple representations to compute orthogonal eigenvectors of symmetric tridiagonal matrices, *Linear Algebra and Applications*, Vol. 387, pp. 1–28, 2004.

[4] Francis, J. G. F.; The QR transformation, *Computer Journal*, Vol. 4, pp 265-271 and pp. 323–345, 1961.

[5] Givens, W.; Numerical computation of the characteristic values of a real symmetric matrix, *Oak Ridge National Lab. Rep.*, ORNL-1574, 1954.

[6] Golub, G. H. and Underwood, R,; The block Lanczos method for computing eigenvalues, *Mathematical Software*, Academic Press, New York, 1977.

[7] Hotelling, H.; Analysis of a complex set of statistical variables into principal components. *Journal of Educational Psychology*, Vol. 24, pp. 417–441, 498–520, 1933.

[8] Komzsik, L.; *The Lanczos Method: Evolution and Application*, SIAM, Philadelphia, 2003.

[9] Lanczos, C.; An iteration method for the solution of the eigenvalue problem of linear differential and integral operators, *J. Res. Nat. Bureau of Standards*, Vol. 45, pp. 255–280, 1950.

[10] Parlett, B. N.; *The Symmetric Eigenvalue Problem*, Prentice-Hall, Englewood Cliffs, NJ, 1980.

[11] Parlett, B. N. and Scott, D. S.; The Lanczos algorithm with selective orthogonalization, *Mathematics of Computation*, Vol. 33, pp. 217–238, 1979.

[12] Rayleigh, L. (J. W. Srutt), On the calculation of the frequency of vibration of system in its gravest mode with an example from hydrodynamics, *Philos. Mag.*, Vol. 47, pp. 556–572, 1899.

[13] Simon, H.; The Lanczos algorithm with partial reorthogonalization, *Mathematics of Computation*, Vol. 42, pp. 115–142, 1984.

[14] Sorensen, D. C.; Implicitly restarted Arnoldi/Lanczos methods for large scale eigenvalue computations, *Parallel Numerical Algorithms*, pp. 119–166, Kluwer, Dordrecht, The Netherlands, 1997.

[15] Wilkinson, J. H.; The calculation of eigenvectors of codiagonal matrices, *Computer Journal*, Vol. 1, pp. 90–96, 1958.

11

Initial value problems

Initial value problems are an important subject of engineering. Many of these problems arise from modeling various physical phenomena, but considerably fewer of them may be solvable analytically, hence the intense interest in approximate solutions. Historical names such as Euler and Taylor are associated with the subject and their classical methods will be discussed in detail.

An excellent survey of the area is in [4]. The focus of this chapter will be initial value problems of ordinary differential equations and systems of ordinary differential equations. Some of the methods are applicable to partial differential equations as well.

It is important to point out that the approximate solution of initial value problems will not be given in the form of an approximate function. Rather, it will be given in terms of approximate solution values at discrete points. This effectively reverses the approach of the first part of the book by seeking a function approximating some discrete points.

11.1   Solution of initial value problems

An initial value problem (IVP) of a first order ordinary differential equation is posed as

$$\frac{dy}{dt} = f(t, y); \; a \leq t \leq b,$$

with the initial condition of

$$y(a) = y_0.$$

The analytic solution of the problem is

$$y(t),$$

which may not be possible or feasible to compute analytically. The solution to such a problem exists under certain conditions. The so-called Lipschitz condition is defined with a constant $L > 0$ on a set $D \in R^2$ as

$$|f(t, y_1) - f(t, y_2)| \le L|y_1 - y_2|,$$

where L is called the Lipschitz constant. If the function $f(t, y)$ is continuous on the domain

$$D = [a \le t \le b; -\infty < y < \infty],$$

and satisfies the Lipschitz condition, then the initial value problem has a unique solution.

The somewhat ambiguous definition of this condition may be improved as follows. Let us define a domain D to be convex if, for any two points in the domain, the line segment connecting the two points is also in the domain in its entirety. If the function is defined and differentiable on a convex domain and there exists a number L such that

$$|\frac{\partial f}{\partial y}(t, y)| \le L$$

for each point in the domain, the function satisfies the Lipschitz condition. This is a more useful and easily verifiable condition, since it essentially verifies that the partial derivative of the function $f(t, y)$ with respect to y is bounded in D.

There is an abundance of methods for the solution of initial value problems, categorized as follows:

In addition to these methods, the initial value problem for systems of differential equations and for higher order equations will also be discussed.

11.2 Single-step methods

We first examine methods that execute one solution step at a time, hence their name.

11.2.1 Euler's method

Euler's method is rarely used in engineering practice, but it is worth discussing here as it defines some concepts of the approximate solution of initial value problems rather intuitively. It is also rather old as its first components appeared in 1738 [2]. The approximate solution is obtained at discrete abscissa locations:

$$t_i = a + ih; i = 0, 1, 2, \ldots, m.$$

The "step size" here is defined as

$$h = \frac{b-a}{m}.$$

Assuming that the solution function is twice differentiable, we write the second order Taylor polynomial in the neighborhood of t_i as

$$y(t) = y(t_i) + y'(t_i)(t - t_i) + y''(\xi_i)\frac{1}{2}(t - t_i)^2,$$

where

$$\xi_i \in (t, t_i).$$

Taking the $t = t_{i+1}$ location and substituting the step size yields

$$y(t_{i+1}) = y(t_i) + hy'(t_i) + \frac{h^2}{2}y''(\xi_i).$$

Based on this, Euler's difference equation method to solve an initial value problem of a first order ordinary differential equation is started from

$$y_0 = y(a),$$

and followed by the iteration steps

$$y_{i+1} = y_i + hf(t, y_i); i = 0, 1, \ldots.$$

The geometric concept of the Euler method, shown in Figure 11.1, is that at every iteration step we approximate the solution curve with a line whose slope is the same as the tangent of the function.

In Figure 11.1 the "epoints" are the exact solution points and "spoints" denote the approximate solution points. The line segments demonstrate the approximate linear solution segments of the Euler method for $f(x)$.

We define the local approximation error as

$$e_{i+1}(h) = \frac{y(t_{i+1}) - y(t_i)}{h} - f(t_i, y_i) = \frac{h}{2} y''(\xi_i),$$

with

$$y''(\xi) \le M; t_i \le \xi \le t_{i+1},$$

$$e_{i+1}(h) = O(h).$$

Euler's method is a stable, but not very accurate method, unless the step size is extremely small.

FIGURE 11.1 Concept of Euler's method

11.2.2   Taylor methods

The Taylor methods use the same concept as the Euler method, but with higher $(2, 3, \ldots)$ order Taylor polynomials. The general nth order Taylor series is of the form

$$y(t_{i+1}) = y(t_i) + hf(t_i, y(t_i)) + \frac{h^2}{2} f'(t_i, y(t_i)) + \ldots \frac{h^n}{n!} f^{(n-1)}(t_i, y(t_i))$$

$$+ \frac{h^{n+1}}{(n+1)!} f^{(n)}(\xi_i, y(\xi_i)).$$

Note that $f(t_i, y(t_i))$ and its derivatives are used in place of the derivatives of $y(t_i)$. We introduce

$$T_n(t_i, y_i) = f(t_i, y_i) + \frac{h}{2} f'(t_i, y_i) + \ldots + \frac{h^{n-1}}{n!} f^{(n-1)}(t_i, y_i).$$

Then the Taylor method may be written as

$$y(t_{i+1}) = y(t_i) + hT_n(t_i, y_i); \ i = 0, 1, \ldots.$$

The local approximation error of a step is

$$e_{i+1}(h) = \frac{y(t_{i+1}) - y(t_i)}{h} - T_n(t_i, y_i) - \frac{h^n}{(n+1)!} f^{(n)}(\xi_i, y(\xi_i)),$$

or

$$e_{i+1}(h) = O(h^n).$$

Therefore, the nth order Taylor method is more accurate than the Euler method. An iteration process for the second order Taylor method, for example, is as follows. Initialize

$$w_0 = y_0,$$

and iterate

$$y_{i+1} = y_i + hT_2(t_i, y_i); i = 0, 1, \ldots$$

or

$$y_{i+1} = y_i + h(f(t_i, y_i) + \frac{h}{2} f'(t_i, y_i)); i = 0, 1, \ldots.$$

Let us now demonstrate these methods with a computational example.

11.2.3   Computational example

We consider the first order differential equation of

$$\frac{dy}{dt} = 1 + \frac{y}{t},$$

with an initial condition of

$$y(1) = 2.$$

We are interested in the solution of $y(3/2)$ and use a step size of $1/4$. For this example,

$$f(t,y) = 1 + \frac{y}{t},$$

and

$$f'(t,y) = \frac{1}{t}.$$

The second order Taylor formula customized for this example is

$$y_{i+1} = y_i + \frac{1}{4}((1 + \frac{y_i}{t_i}) + \frac{1}{8}\frac{1}{t_i}).$$

Substituting the initial condition yields

$$y_1 = 2 + \frac{1}{4}((1 + \frac{2}{1}) + \frac{1}{8}\frac{1}{1}) = \frac{89}{32},$$

and

$$y_2 = \frac{89}{32} + \frac{1}{4}((1 + \frac{89/32}{5/4}) + \frac{1}{8}\frac{4}{5}) = \frac{578}{160} = 3.6125.$$

The analytic solution of the differential equation is

$$y = t \cdot ln(t) + 2t + c,$$

with the initial condition forcing $c = 0$. The exact solution value at $t = 3/2$ is

$$y(3/2) = 3.60819.$$

Assuming, reasonably, that M, the upper bound for $y''(\xi)$, is unity, the local approximation error at each step has an upper bound of

$$\iota_1(1/4) \leq \frac{h^2}{3!} = \frac{1}{96} = 0.0104.$$

The difference between the approximate and the exact result is

$$0.00431,$$

TABLE 11.1

Taylor method example

i	t_i	y_i	$y(t_i)$	$y_i - y(t_i)$
0	1	2	2	
1	1.25	2.78125	2.77893	0.00232
2	1.50	3.61250	3.60819	0.00431
3	1.75	4.48542	4.47933	0.00609
4	2.00	5.39400	5.38629	0.00771

well within the expected range. Two more steps are executed and all are sum-marized in Table 11.1, where the third column is the exact result.

Note that the error between the exact and the approximate result shown in the last column of Table 11.1 is gradually increasing, in accordance with the fact that each step produces a distinct approximation error and they are being accumulated.

A disadvantage of the Taylor methods is the need to compute the derivatives for the Taylor polynomial analytically, a feat that may not be accomplished in practical engineering problems. The next class of methods is aimed at over-coming this limitation.

11.2.4   Runge–Kutta methods

The methods in this class [10] compute the derivatives only approximately, hence they are preferred and more practical in engineering computations. Just as in the Taylor methods, various order formulations are available. Note that these methods are still of the single-step family.

For the sake of clarity, we first restrict our attention to the $n = 2$ case. The Taylor method of order $n = 2$ was written as

$$y_{i+1} = y_i + hT_2(t_i, y_i); i = 0, 1, \ldots,$$

with

$$T_2(t, y) = f(t, y) + \frac{h}{2}f'(t, y).$$

Using the chain rule, the derivative on the right-hand side is

$$f'(t, y) = \frac{df}{dt}(t, y) = \frac{\partial f}{\partial t}(t, y) + \frac{\partial f}{\partial y}(t, y)y'(t)$$

and

$$T_2(t,y) = f(t,y) + \frac{h}{2}\frac{\partial f}{\partial t}(t,y) + \frac{h}{2}\frac{\partial f}{\partial y}(t,y)f(t,y).$$

The fundamental idea of the Runge–Kutta method is to approximate the derivatives needed in this expression. We assume an approximation of $f(t,y)$ with its two-dimensional, first order Taylor polynomial (not to be confused with Taylor's method) as

$$f(t+k_1, y+k_2) = f(t,y) + k_1\frac{\partial f}{\partial t}(t,y) + k_2\frac{\partial f}{\partial y}(t,y)f(t,y) + R_1(t+k_1, y+k_2).$$

Here the remainder term is

$$R_1(t+k_1, y+k_2) = \frac{k_1^2}{2}\frac{\partial^2 f}{\partial t^2}(\xi,\zeta) + k_1k_2\frac{\partial^2 f}{\partial t\partial y}(\xi,\zeta) + \frac{k_2^2}{2}\frac{\partial^2 f}{\partial y^2}(\xi,\zeta),$$

where

$$\xi \in (t, t+k_1); \zeta \in (y, y+k_2).$$

Matching terms between the last two equations, we obtain

$$k_1 = \frac{h}{2},$$

and

$$k_2 = \frac{h}{2}f(t,y).$$

Hence the approximation of $T_2(t,y)$ is

$$T_2(t,y) = f(t+\frac{h}{2}, y+\frac{h}{2}f(t,y)),$$

and the Runge–Kutta method of order 2 is formed as

$$y_{i+1} = y_i + hf(t_i + \frac{h}{2}, y_i + \frac{h}{2}f(t_i, y_i)); i = 0, 1, \ldots .$$

The initialization steps of

$$k_1 = \frac{h}{2},$$

and

$$k_2 = \frac{h}{2}f(t_i, y_i),$$

followed by an iteration step of

$$y_{i+1} = y_i + hf(t_i + k_1, y_i + k_2); i = 0, 1, \ldots$$

constitute the algorithm executing the method. Based on R_1 and $k_1 = h/2$ it can be seen that the method's local error is still $O(h^2)$, which is very good considering the work saved by not computing the analytic derivatives is tremendous and sometimes makes the difference between whether a problem may be solved or not.

The also well-used fourth order Runge–Kutta method is based on the same concept and formulated as

$$k_1 = hf(t_i, y_i),$$

$$k_2 = hf(t_i + \frac{h}{2}, y_i + \frac{k_1}{2}),$$

$$k_3 = hf(t_i + \frac{h}{2}, y_i + \frac{k_2}{2}),$$

$$k_4 = hf(t_{i+1}, y_i + k_3),$$

and

$$y_{i+1} = y_i + \frac{1}{6}(k_1 + 2k_2 + 2k_3 + k_4).$$

This method has an $O(h^4)$ local approximation error.

11.2.5 Computational example

We revisit the first order differential equation of

$$\frac{dy}{dt} = 1 + \frac{y}{t},$$

with an initial condition of

$$y_0 = y(t_0) = y(1) = 2.$$

We are going to execute one iteration of the Runge–Kutta method with a step size of $1/4$. The method is executed in the following algorithmic steps:

$$k_1 = \frac{h}{2} = \frac{1}{8}.$$

For $i = 0$,

$$k_2 = \frac{h}{2}f(t_0, y_0) = \frac{1}{8}(1 + \frac{y_0}{t_0}) = \frac{1}{8}(1 + \frac{2}{1}) = \frac{3}{8}$$

and

$$y_1 = y_0 + hf(t_0 + k_1, y_0 + k_2) = 2 + \frac{1}{4}(1 + \frac{2 + 3/8}{1 + 1/8}) = 2.77778.$$

This value is in very good agreement with the analytical solution of 2.77893 shown in Table 11.1. This is, indeed, a testament to the practical properties of the Runge–Kutta method.

11.3   Multistep methods

The multistep methods execute more than one step of the iterations at a time. Depending on whether the step size is fixed or variable, we have two main groups.

The fixed step size m-step multistep methods may be written in the generic form of

$$y_{i+1} = a_{m-1}y_i + a_{m-2}y_{i-1} + \ldots + a_0 y_{i+1-m}$$
$$+ h(b_m f(t_{i+1}, y_{i+1}) + b_{m-1}f(t_i, y_i) + \ldots + b_0 f(t_{i+1-m}, y_{i+1-m})).$$

Here we define

$$h = \frac{t_n - t_0}{n}.$$

Depending on the value of b_m, i.e., whether we use the new iterate value, we may have implicit ($b_m \neq 0$) or explicit ($b_m = 0$) methods. These methods are also sometimes called closed and open, respectively.

Naturally, for an m-step method m initial conditions must be supplied, i.e.,

$$y_0, y_1, \ldots, y_{m-1}$$

must be given to start. These values may be obtained by any of the single-step methods. The methods of this class are called Adams methods.

11.3.1   Explicit methods

For an explicit Adams method, consider our problem:

$$y' = f(t, y)$$

and integrate formally

$$\int_{t_i}^{t_{i+1}} y'(t)dt = y(t)|_{t_i}^{t_{i+1}} = y(t_{i+1}) - y(t_i) = \int_{t_i}^{t_{i+1}} f(t, y(t))dt.$$

Hence

$$y(t_{i+1}) = y(t_i) + \int_{t_i}^{t_{i+1}} f(t, y(t))dt.$$

Of course we cannot compute this integral, since the inner function $y(t)$ is unknown, so we approximate the integrand with Newton's equidistant backward difference formula of

$$f(t, y(t)) = \sum_{k=0}^{m-1} (-1)^k \binom{-s}{k} \nabla^k f(t_i, y(t_i)) + R_m.$$

Following the Chapter 1 discussion on Newton's equidistant backward formulae,

$$\nabla f(t_i, y(t_i)) = f(t_i, y(t_i)) - f(t_{i-1}, y(t_{i-1})),$$

and

$$\nabla^k f(t_i, y(t_i)) = \nabla \nabla^{k-1} f(t_i, y(t_i)); k = 1, 2, \ldots.$$

The R_m error term of the approximation of the function is

$$R_m = \frac{f^{(m)}(\xi_i, y(\xi_i))}{m!}(t - t_i)(t - t_{i-1})(t - t_{i-2}) \ldots (t - t_{i-(m-1)}).$$

Introducing

$$t - t_i + s \cdot h,$$

and

$$dt = h \, ds,$$

the integral becomes

$$\int_{t_i}^{t_{i+1}} f(t, y(t))dt = \sum_{k=0}^{m-1} h \nabla^k f(t_i, y(t_i)) I_k + \int_{t_i}^{t_{i+1}} R_m dt,$$

where

$$I_k = (-1)^k \int_0^1 \binom{-s}{k} ds.$$

These integrals may be easily evaluated for various k values. For our discussion we need the integrals for $k = 0, 1, 2, 3$ and they are

$$I_0 = 1, I_1 = \frac{1}{2}, I_2 = \frac{5}{12}, I_3 = \frac{3}{8}.$$

Substituting these values, we obtain

$$y(t_{i+1}) = y(t_i) + h(f(t_i, y(t_i)) + \frac{1}{2}\nabla f(t_i, y(t_i)) + \frac{5}{12}\nabla^2 f(t_i, y(t_i))$$

$$+ \frac{3}{8}\nabla^3 f(t_i, y(t_i)) + \ldots) + \int_{t_i}^{t_{i+1}} R_m(t_i, y(t_i))dt,$$

which is the basis of the Adams methods.

Let us now focus on the error of this approximation. Using the substitutions as above, the error term of the integration is

$$\int_{t_i}^{t_{i+1}} R_m dt = \frac{h^{(m+1)}}{m!} \int_0^1 s(s+1)\ldots(s+m-1)f^{(m)}(\xi_i, y(\xi_i))ds$$

$$= \frac{h^{(m+1)}}{m!} f^{(m)}(\zeta_i, y(\zeta_i)) \int_0^1 s(s+1)\ldots(s+m-1)ds$$

$$= \frac{h^{(m+1)}}{m!} f^{(m)}(\zeta_i, y(\zeta_i))(-1)^m \int_0^1 \binom{-s}{m} ds.$$

Computing the above for $m = 3$ yields

$$y(t_{i+1}) = y(t_i) + h(f(t_i, y(t_i)) + \frac{1}{2}\nabla f(t_i, y(t_i)) + \frac{5}{12}\nabla^2 f(t_i, y(t_i)))$$

$$= y(t_i) + h(f(t_i, y(t_i)) + \frac{1}{2}(f(t_i, y(t_i)) - f(t_{i-1}, y(t_{i-1})))$$

$$+ \frac{5}{12}(f(t_i, y(t_i)) - 2f(t_{i-1}, y(t_{i-1})) + f(t_{i-2}, y(t_{i-2}))))$$

$$= y(t_i) + \frac{h}{12}(23f(t_i, y(t_i)) - 16f(t_{i-1}, y(t_{i-1})) + 5f(t_{i-2}, y(t_{i-2}))).$$

Based on the above, the Adams–Bashworth three-step explicit method is

$$y_{i+1} = y_i + \frac{h}{12}(23f(t_i, y_i) - 16f(t_{i-1}, y_{i-1}) + 5f(t_{i-2}, y_{i-2})),$$

for $i = 2, 3, \ldots$ with y_2, y_1 and y_0 initial conditions. The specific error term is computed as

$$h^4 f^{(3)}(\zeta_i, y(\zeta_i))(-1)^3 \int_0^1 \binom{-s}{3} ds = \frac{3h^4}{8} f^{(3)}(\xi_i, y(\xi_i)).$$

The local error is obtained after division by the step size

$$e_{i+1}(h) = \frac{3h^3}{8} y^4(\xi_i).$$

Here

$$\xi_i, \zeta_i \in (t_{i-2}, t_{i+1}).$$

The local error of the three-step Adams–Bashworth method is of $O(h^3)$.

11.3.2 Implicit methods

In order to provide an implicit method along the same lines, another integration term of

$$\int_{t_i}^{t_{i+1}} f(t, y(t))dt$$

needs to be included. This process will give rise to the Adams–Moulton [7] methods, of which the three-step formula is listed here.

$$y_{i+1} = y_i + \frac{h}{24}(9f(t_{i+1}, y_{i+1}) + 19f(t_i, y_i) - 5f(t_{i-1}, y_{i-1}) + f(t_{i-2}, y_{i-2})),$$

for $i = 2, 3, \ldots$ with y_2, y_1 and y_0 initial conditions. The local error is

$$e_{i+1}(h) = -\frac{19h^3}{720}y^5(\xi_i).$$

This error term is smaller than the error term of the corresponding explicit Adams–Bashworth formula, a fact compensating for the additional complexity of the formula. The problem is that it is an implicit equation that has the y_{i+1} term on both sides requiring a solution step, instead of a substitution only.

11.3.3 Predictor–corrector technique

In order to overcome the difficulty of the solution step of the implicit methods, methods to combine the explicit and implicit techniques gained ground. They are called predictor–corrector methods.

The idea is to execute an explicit step to obtain an approximation for y_{i+1} denoted by y_{i+1}^p, for prediction. Then an implicit solution step is executed using this predicted value on the right-hand side to obtain a corrected approximation for y_{i+1}.

Let us consider the explicit Adams–Bashworth two-step formula of

$$y_{i+1}^p = y_i + \frac{h}{2}(3f(t_i, y_i) - f(t_{i-1}, y_{i-1}))$$

as a predictor step. The two-step implicit Adams–Moulton formula of

$$y_{i+1} = y_i + \frac{h}{12}(5f(t_{i+1}, y_{i+1}^p) + 8f(t_i, y_i) - f(t_{i-1}, y_{i-1}))$$

may be the accompanying corrector step.

11.3.3.1   Computational example

To demonstrate the power of this technique, we again revisit the earlier example of this chapter:

$$\frac{dy}{dt} = 1 + \frac{y}{t}.$$

We will still use a step size of $1/4$ and use the two-step predictor–corrector method. The original initial condition was

$$y_0 = y(t_0) = y(1) = 2.$$

We will use the analytic solution as the required second initial condition for simplicity.

$$y_1 = y(t_1) = y(1.25) = 2.77893.$$

The customized Adams–Bashworth two-step formula for this problem is

$$y_{i+1} = y_i + \frac{1}{8}\left(3\left(1 + \frac{y_i}{t_i}\right) - \left(1 + \frac{y_{i-1}}{t_{i-1}}\right)\right) = y_i + \frac{1}{4} + \frac{3}{8}\frac{y_i}{t_i} - \frac{1}{8}\frac{y_{i-1}}{t_{i-1}}.$$

The predicted value is

$$y_2^p = 3.61261.$$

The customized Adams–Moulton two-step formula is

$$y_{i+1} = y_i + \frac{5}{48}\left(1 + \frac{y_{i+1}^p}{t_{i+1}}\right) + \frac{8}{48}\left(1 + \frac{y_i}{t_i}\right) - \frac{1}{48}\left(1 + \frac{y_{i-1}}{t_{i-1}}\right).$$

This yields

$$y_{i+1} = 3.60866,$$

which is an order of magnitude better than the predictor step. The errors of some of the methods discussed in this chapter and applied to this problem are shown in Table 11.2.

TABLE 11.2
Error of second order methods

Method	$y(t_2) - y_2$
Taylor method	0.00431
Runge–Kutta	0.00736
Adams–Bashworth	0.00442
Predictor–corrector	0.00047

The techniques discussed so far have all used a constant step size. Additional advantage may be gained by employing a variable step size. Such methods, furthermore, employ a variation of the earlier methods in certain combinations.

11.3.4 Gragg's method of extrapolation

As the name indicates, this method has some flavors of Richardson's extrapolation [5] in it. Gragg's method employs several intermediate steps. Consider the initial value problem posed as

$$y' = f(t, y)$$

with an initial condition of

$$y(a) = y(t_0) = y_0$$

given and the solution value $y(b)$ sought. In the first intermediate step Gragg's method executes a single step of Euler's method with a half step size of

$$h_0 = \frac{b - a}{2}$$

as

$$\overline{y}_{1/2} = y_0 + h_0 f(t_0, y_0).$$

The subscript $1/2$ indicates that this approximation is made for a half step and the bar in this section indicates intermediate results. Then a step of the second order Runge–Kutta method, also known as the midpoint method, is executed as

$$\overline{y}_1 = y_0 + hf(t_0 + h_0, y_0 + h_0 f(t_0, y_0)) = y_0 + 2h_0 f(t_0 + h_0, \overline{y}_{1/2}).$$

Here the right-hand side was obtained by substituting the "half" step size Euler approximation since

$$h = 2h_0 = b - a.$$

We compute the average of the two values as

$$\overline{y}_{3/4} = \frac{\overline{y}_1 + \overline{y}_{1/2}}{2}.$$

We fit a line through the points

$$(b, y_1)$$

and

$$(t_{3/4}, \overline{y}_{3/4}),$$

where

$$t_{3/4} = t_0 + \frac{3}{4}h = t_0 + \frac{3}{2}h_0.$$

The equation of the line is

$$y - \overline{y}_{3/4} = f(b, \overline{y}_1)(t - t_{3/4}).$$

The slope used is the slope of the function at b. Substituting $t = b$, we extrapolate to find an equation for the improved approximate solution.

$$y_1 = \frac{1}{2}(\overline{y}_{1/2} + \overline{y}_1 + h_0 f(b, \overline{y}_1)).$$

The solution of this equation provides the extrapolation. The scheme is shown graphically in Figure 11.2 with E, M, G points denoting the $\overline{y}_{1/2}, \overline{y}_1, y_1$ points, obtained by the Euler, midpoint and Gragg's steps, respectively.

FIGURE 11.2 Gragg's extrapolation scheme

The algorithmic process is summarized below.

Compute step size:

$$h_0 - \frac{b-a}{2}.$$

Euler step:

$$\overline{y}_{1/2} = y_0 + h_0 f(t_0, y_0).$$

Midpoint step:

$$\overline{y}_1 = y_0 + 2h_0 f(t_0 + h_0, \overline{y}_{1/2}).$$

Extrapolation step:

$$y_1 = \frac{1}{2}(\overline{y}_{1/2} + \overline{y}_1 + h_0 f(b, \overline{y}_1)).$$

This process may be generalized into a multilevel extrapolation technique based on systematically different subdivided step sizes. Such step size sequences may be obtained as

$$h_j = \frac{h}{q_j}; q_j = 2^{j+1}; j = 0, 1, \ldots.$$

Hence

$$h_0 = \frac{h}{q_0}, h_1 = \frac{h}{q_1}, h_2 = \frac{h}{q_2}, \ldots.$$

We introduce the notation $y_{1,j}$, where the second subscript represents the step size index, such that $y_{1,0}$ is the above computed y_1 with step size h_0.

In the second level of the scheme with a step size of

$$h_1 = \frac{h_0}{2} = \frac{h}{4},$$

the $y_{1,1}$ approximation is computed by executing two midpoint (Runge–Kutta) steps following the Euler step.

Euler step:

$$\overline{y}_{1/4} = y_0 + h_1 f(t_0, y_0).$$

Midpoint step 1:

$$\overline{y}_{1/2} = y_0 + 2h_1 f(t_0 + h_1, \overline{y}_{1/4}).$$

Midpoint step 2:

$$\overline{y}_{3/4} = \overline{y}_{1/4} + 2h_1 f(t_0 + 2h_1, \overline{y}_{1/2}).$$

Midpoint step 3:

$$\overline{y}_1 = \overline{y}_{1/2} + 2h_1 f(t_0 + 3h_1, \overline{y}_{3/4}).$$

Extrapolation step:

$$y_{1,1} = \frac{1}{2}(\overline{y}_{3/4} + \overline{y}_1 + h_1 f(b, \overline{y}_1)).$$

The seemingly increasing difficulty is rewarded handsomely in accuracy. Since the local error of the Runge–Kutta second order (the midpoint) steps is $O(h^2)$, for the second level the error is

$$O(h_1^2) = O(\frac{h_0}{4}^2) = O(\frac{h^2}{16}),$$

which is an order of magnitude better. Furthermore, the error may be reduced as follows. For the first level,

$$y(b) = y(a + h) = y_{1,0} + \epsilon_1 h_0^2 + \epsilon_2 h_0^4 + \ldots,$$

and for the second level,

$$y(b) = y(a + h) = y_{1,1} + \epsilon_1 h_1^2 + \epsilon_2 h_1^4 + \ldots.$$

Here the ϵ are the constant parts of the error terms. The two approximations in common terms may be compared as

$$y(b) = y(a + h) = y_{1,0} + \epsilon_1 \frac{h^2}{4} + \epsilon_1 \frac{h^4}{16} + \ldots,$$

and

$$y(b) = y(a + h) = y_{1,1} + \epsilon_1 \frac{h^2}{16} + \epsilon_1 \frac{h^4}{256} + \ldots.$$

Appropriate multiplication and subtraction (utilizing the technique introduced by Romberg and discussed in Section 6.2.2) yields

$$y_1 = y_{1,1} + \frac{1}{3}(y_{1,1} - y_{1,0}),$$

which is a further improvement on the approximation.

Gragg's method can be extended to further levels, but these may reach a point of diminishing returns. Certainly three levels may still be useful.

11.3.5   Computational example

The complexity of Gragg's method definitely warrants an example, although the technique is almost too laborious for hand computation. The technique, of course, is a prime candidate for computer implementation.

The example of

$$y' = 1 + \frac{y}{t}$$

pervading this chapter is used again with the initial condition

$$y(a) = y(1) = 2,$$

seeking the solution of

$$y(b) = y(5/4).$$

The respective step sizes are

$$h = \frac{1}{4}; h_0 = \frac{1}{8}; h_1 = \frac{1}{16}.$$

The execution of two levels of Gragg's extrapolation process follows.

First level with $h_0 = \frac{1}{8}$:

Euler step:

$$\overline{y}_{1/2} = 2 + \frac{1}{8}(1 + \frac{2}{1}) = \frac{19}{8}.$$

Midpoint step:

$$\overline{y}_1 = 2 + 2(\frac{1}{8})(1 + \frac{19/8}{9/8}) = \frac{25}{9}.$$

Extrapolation step:

$$y_{1,0} = \frac{1}{2}(\frac{19}{8} + \frac{25}{9}\frac{1}{8}(1 + \frac{25/9}{5/4})) = 2.77778.$$

Second level with $h_1 = \frac{1}{16}$:

Euler step:

$$\overline{y}_{1/4} = 2 + \frac{1}{16}(1 + \frac{2}{1}) = \frac{35}{16}.$$

Midpoint step 1:

$$\overline{y}_{1/2} = 2 + 2(\frac{1}{16})(1 + \frac{35/16}{17/16}) = \frac{81}{34}.$$

Midpoint step 2:

$$\overline{y}_{3/4} = \frac{35}{16} + 2(\frac{1}{16})(1 + \frac{81/34}{9/8}) = \frac{701}{272}.$$

Midpoint step 3:

$$\overline{y}_1 = \frac{81}{34} + 2(\frac{1}{16})(1 + \frac{701/272}{19/16}) = \frac{1795}{646}.$$

Extrapolation step:

$$y_{1,1} = \frac{1}{2}(\frac{701}{272} + \frac{1795}{646} + \frac{1}{16}(1 + \frac{1795/646}{5/4})) = 2.77864.$$

Finally the Romberg's method style refinement is computed as

$$y_1 = 2.77864 + \frac{1}{3}(2.77864 - 2.77778) = 2.77892.$$

The increased accuracy of these steps is demonstrated in Table 11.3 in comparison with the exact solution of $y(5/4) = 2.77893$.

TABLE 11.3
Error of Gragg's method

Step	$y(t_1) - y_1$
First level	0.00115
Second level	0.00029
Refinement	0.00001

The refined final value only differs from the exact value in the last digit with an error of $O(10^{-5})$.

11.3.6 Fehlberg's method of step size adjustment

Another way to improve on the approximation by varying the step size is the method of Fehlberg [3]. It simultaneously executes two sequences of Taylor methods, one with order m and another one with order $m + 1$. The approximations are

$$y_{i+1} = y_i + hT_m + O(h^{m+1}),$$

and

$$\overline{y}_{i+1} = \overline{y}_i + hT_{m+1} + O(h^{m+2}).$$

Both of these are, of course, approximating the exact solution:

$$y(t_{i+1}) \approx y_{i+1} \approx \overline{y}_{i+1}.$$

The local errors of the two approximations are

$$e_{i+1}(h) = \frac{y(t_{i+1}) - y_{i+1}}{h},$$

and

$$\overline{e}_{i+1}(h) = \frac{y(t_{i+1}) - \overline{y}_{i+1}}{h}.$$

With some algebraic manipulations we can express one in terms of the other:

$$e_{i+1}(h) = \frac{(y(t_{i+1}) - \overline{y}_{i+1}) + (\overline{y}_{i+1} - y_{i+1})}{h} = \overline{e}_{i+1}(h) + \frac{y_{i+1} - y_{i+1}}{h}.$$

Assuming that the local truncation error of the higher order method is much less, the approximate local error of the lower order method is computed as

$$e_{i+1}(h) \approx \frac{\overline{y}_{i+1} - y_{i+1}}{h}.$$

Note the significant distinction here: this is not a bound for the local error, but an approximation for it. This enables the Fehlberg method of step size adjustment as follows. At any step of the approximation process, when the estimated local error exceeds a certain value,

$$e_{i+1}(h) \geq \epsilon,$$

the step size is adjusted (i.e., reduced) with a multiplier α,

$$\overline{h} = \alpha h.$$

The estimated local error with the adjusted step size is

$$e_{i+1}(\alpha h) = O(\alpha h)^m \approx k(\alpha h)^m = \alpha^m (k \cdot h^m) = \alpha^m e_{i+1}(h).$$

Enforcing the adjusted error to be within the requirement

$$e_{i+1}(\alpha h) \leq \epsilon$$

and substituting the original step size error yields the value of adjustment,

$$\alpha = \left(\frac{\epsilon h}{|\overline{y}_{i+1} - y_{i+1}|}\right)^{1/m}.$$

The process then continues with the new step size resulting in a variable step size method. The method is used in practice in connection with higher order Runge–Kutta methods. Often a fourth order method is executed in connection with a fifth order to estimate the error and adjust appropriately. The application is more practical in computer implementation and less for manual computations.

On a final note for the variable step size methods, we mention that the predictor–corrector technique is sometimes used in different steps in the predictor and the corrector phase, hence qualifying for being a variable step size method. The combination of a three-step explicit Adams–Bashworth predictor step with a two-step implicit Adams–Moulton correction is used often in practical implementations.

11.3.7   Stability of multistep techniques

An approximation method is considered to be stable if small changes in the initial conditions result in small changes in the approximations. The stability of a multistep method is evaluated as follows.

Consider an m-step method of

$$y_{i+1} = a_{m-1}y_i + a_{m-2}y_{i-1} + \ldots + a_0 y_{i-m+1}.$$

The characteristic polynomial of such a method is

$$p(\lambda) = \lambda^m - a_{m-1}\lambda^{m-1} - \ldots - a_1\lambda - a_0 = 0.$$

If all the roots of the characteristic equation satisfy

$$|\lambda_i| \leq 1; i = 1, 2, \ldots m,$$

then the method is called stable. Furthermore, a method is called strongly stable if only one root satisfies the equality.

11.4   Initial value problems of systems of ordinary differential equations

We now consider the initial value problem of a system of n linear ordinary differential equations (ODEs).

$$\frac{dy_1}{dt} = f_1(t, y_1, y_2, \ldots, y_n),$$

$$\frac{dy_2}{dt} = f_2(t, y_1, y_2, \ldots, y_n),$$

and

$$\frac{dy_n}{dt} = f_n(t, y_1, y_2, \ldots, y_n).$$

The system has n initial conditions:

$$y_1(t_0) = y_{1,0},$$

$$y_2(t_0) = y_{2,0},$$

and

$$y_n(t_0) = y_{n,0}.$$

Naturally, we seek n solutions of

$$y_1 = y_1(t),$$

$$y_2 = y_2(t),$$

and

$$y_n = y_n(t).$$

The Lipschitz condition for a function of n variables is

$$|f(t, y_1, y_2, \ldots, y_n) - f(t, w_1, w_2, \ldots, w_n)| \leq L \sum_{k=1}^{n} |y_k - w_k|,$$

where the constant L is the Lipschitz constant. We define a convex domain

$$D = [(t, y_1, y_2, \ldots, y_n); a \leq t \leq b; -\infty < y_k < \infty, k = 1, 2, \ldots, n].$$

If the first partial derivatives are continuous on the domain and satisfy

$$\frac{\partial f(t, y_1, y_2, \ldots, y_n)}{\partial y_i} \leq L,$$

then the function satisfies the Lipschitz condition and the system has a unique solution.

11.4.1   Picard's successive approximation

The successive approximation solution for the first initial value problem posed in Section 11.1 was originally proposed by the earlier mentioned and referenced Picard. For the problem

$$\frac{dy}{dt} = f(t, y); y(t_0) = y_0,$$

where the function f satisfies the Lipschitz condition, the iterative steps $k = 1, 2, \ldots$ produce the sequence of functions

$$y_k = y_0 + \int_{t_0}^t f(\xi, y_{k-1}(\xi))d\xi.$$

This series of functions converges to the unique solution of the initial value problem.

Turning our attention to the system of equations case, assuming that all the $f_i, i = 1, \ldots, n$ functions satisfy their respective Lipschitz conditions and are bounded in the domain D as stated in the preceding section, the Picard solution may be generalized as

$$y_{1,k} = y_{1,0} + \int_{t_0}^t f_1(\xi, y_{1,k-1}(\xi), \ldots, y_{n,k-1}(\xi))d\xi,$$

$$y_{2,k} = y_{2,0} + \int_{t_0}^t f_2(\xi, y_{1,k-1}(\xi), \ldots, y_{n,k-1}(\xi))d\xi,$$

and

$$y_{n,k} = y_{n,0} + \int_{t_0}^t f_n(\xi, y_{1,k-1}(\xi), \ldots, y_{n,k-1}(\xi))d\xi,$$

where the second subscript (k) is the iteration number and $y_{i,0} = y_i(t_0)$.

Assuming that all functions in the system are continuous and bounded, the successive approximation method produces a series converging to the solution [1]. The evaluation of this is certainly far from trivial, but in some cases (especially when the functions of the system are constants) may be easily computed. An example will be shown in the section after the next to illuminate the method.

11.4.2   Runge–Kutta solution

The solution for the current problem may also be attained by a generalization of the Runge–Kutta method. While this may be based on any order of the method, we will use the second order method for simplicity of the introduction. Recalling Section 11.2.4 and considering the jth equation, we write a step as

$$y_{j,i+1} = y_{j,i} + hf_j(t_i + \frac{h}{2}, y_{j,i} + \frac{h}{2}f(t_i, y_{j,i})),$$

for $i = 0, 1, 2, \ldots$. Here the double subscript is interpreted as follows: the second subscript is the iteration counter as before, and the first subscript specifies one of the n linear ODEs.

In the case of a system of equations, $j = 1, 2, \ldots, n$, the computation is presented in the following algorithm.

For $i = 0, 1, 2, \ldots$

For $j = 1, 2, \ldots, n$ compute all

$$k_{j,i} = \frac{h}{2} f_j(t_i, y_{1,i}, y_{2,i}, \ldots, y_{n,i}).$$

For $j = 1, 2, \ldots, n$ compute all

$$l_{j,i} = h f_j(t_i + \frac{h}{2}, y_{1,i} + k_{1,i}, y_{2,i} + k_{2,i}, \ldots, y_{n,i} + k_{n,i}).$$

For $j = 1, 2, \ldots, n$ compute all

$$y_{j,i+1} = y_{j,i} + l_{j,i}.$$

End of loop on i.

The algorithm has an external loop over the iteration process and three internal loops over the system of equations.

11.4.3 Computational example

We consider the very simple, but demonstrative, example of the system

$$\frac{dy_1}{dt} = f_1(t, y_1, y_2) = y_2,$$

and

$$\frac{dy_2}{dt} = f_2(t, y_1, y_2) = -y_1 + 2y_2.$$

The initial conditions are

$$y_1(0) = y_{1,0} = 1,$$

and

$$y_2(0) = y_{2,0} = 2.$$

The solution at $t = \frac{1}{2}$ is sought.

Let us first consider the successive approximation method of Picard. The first iteration is as follows:

$$y_{1,1} = y_{1,0} + \int_0^t y_{2,0} d\xi = 1 + \int_0^t 2 d\xi = 1 + 2t,$$

and

$$y_{2,1} = y_{2,0} + \int_0^t (2y_{2,0} - y_{1,0}) d\xi = 2 + 3t.$$

Subsequently

$$y_{1,2} = 1 + \int_0^t 2 + 3\xi d\xi = 1 + 2t + \frac{3}{2} t^2,$$

and

$$y_{2,2} = 2 + \int_0^t (2(2 + 3\xi) - (1 + 2\xi)) d\xi = 2 + 3t + 2t^2.$$

A very visible pattern emerges: each iteration produces an additional term in the power series. The results of the next two steps are

$$y_{1,3} = 1 + 2t + \frac{3}{2} t^2 + \frac{2}{3} t^3,$$

$$y_{2,3} = 2 + 3t + 2t^2 + \frac{5}{6} t^3,$$

and finally

$$y_{1,4} = 1 + 2t + \frac{3}{2} t^2 + \frac{2}{3} t^3 + \frac{5}{24} t^4,$$

$$y_{2,4} = 2 + 3t + 2t^2 + \frac{5}{6} t^3 + \frac{1}{4} t^4.$$

Substituting the desired location of the result ($t = 1/2$) produces the approximate solution function values shown in Table 11.4.

TABLE 11.4
Result of Picard's
method

Iteration	y_1	y_2
1	2.0	3.5
2	2.375	4.0
3	2.458	4.104
4	2.471	4.119

One numerical step of the Runge–Kutta process is executed below. The solution at $t = \frac{1}{2}$ is sought, i.e., $h = \frac{1}{2}$.

$i = 0$:

$j = 1$:

$$k_{1,0} = \frac{1}{4}2 = \frac{1}{2}.$$

$j = 2$:

$$k_{2,0} = \frac{1}{4}(-1 + 2 \cdot 2) = \frac{3}{4}.$$

$j = 1$:

$$l_{1,0} = \frac{1}{2}(2 + \frac{3}{4}) = \frac{11}{8}.$$

$j = 2$:

$$l_{2,0} = \frac{1}{2}(-(1 + \frac{1}{2}) + 2(2 + \frac{3}{4})) - 2.$$

$j = 1$:

$$y_{1,1} = 1 + \frac{11}{8} = \frac{19}{8} = 2.375.$$

$j = 2$:

$$y_{2,1} = 2 + 2 - 4.$$

Interestingly this is exactly the same solution obtained above with two steps of the successive approximation technique, which is not a surprise because we used a second order Runge–Kutta method.

The system's analytic solution is

$$y_1(t) = e^t + te^t,$$

and

$$y_2(t) = 2e^t + te^t,$$

yielding the above initial conditions at $t = 0$. The exact solution at $t = \frac{1}{2}$ is

$$y_1(\frac{1}{2}) = \sqrt{e} + \frac{\sqrt{e}}{2} = 2.4731,$$

and

$$y_2(\frac{1}{2}) = 2\sqrt{e} + \frac{\sqrt{e}}{2} = 4.1218.$$

The results are moderately accurate because the step size chosen for easy computation is rather large for the particular functions in the initial neighborhood.

Nevertheless, both methods were converging toward the exact solution.

On the other hand, the Picard solution was producing a series of power functions that asymptotically can be used to describe the analytic solution. Since the Taylor series of the e^x function around zero is

$$e^x = 1 + x + \frac{x^2}{2!} + \frac{x^3}{3!} + \frac{x^4}{4!} + \dots$$

the series expansion of the analytic solutions is

$$e^x + xe^x = 1 + 2x + \frac{3}{2}x^2 + \frac{2}{3}x^3 + \frac{5}{24}x^4 + \dots$$

and

$$2e^x + xe^x = 2 + 3x + 2x^2 + \frac{5}{6}x^3 + \frac{1}{4}x^4 + \dots \ .$$

These solutions are identical to the fourth step of the successive approximation solution.

11.5 Initial value problems of higher order ordinary differential equations

Initial value problems (IVPs) of higher order differential equations (ODEs) may also be solved with a similar approach. A general nth order ODE initial value problem is posed as

$$\frac{d^n y(t)}{dt^n} = g(t) + a_0 y(t) + a_1 y'(t) + a_2 y''(t) + \dots a_{n-1} y^{(n-1)}(t),$$

with initial conditions

$$y_1(t_0) = y_{0,0},$$
$$y'(t_0) = y_{1,0},$$
$$y''(t_0) = y_{2,0},$$

and

$$y^{(n-1)}(t_0) = y_{n-1,0}.$$

The solution of this problem is obtained by reformulating it as an initial value problem of a system of linear ODEs. We develop the method while focusing on the $n = 2$ case,

$$y''(t) = g(t) + a_0 y(t) + a_1 y'(t) = f(t, y, y').$$

We assign new variables as follows:

$$w_1(t) = y(t),$$
$$w_2(t) = y'(t).$$

Since then

$$w_1'(t) = w_2,$$

and

$$w_2'(t) = y''(t),$$

the converted system of linear ODEs is

$$\frac{dw_1(t)}{dt} = y'(t) = w_2,$$

and

$$\frac{dw_2(t)}{dt} = f(t, y, y') = f(t, w_1, w_2).$$

This system may be solved by the method in the last section and the approximate solution is

$$y_i = w_{1,i}.$$

The solution scheme is quite useful in engineering applications, where second order differential equations occur frequently, but the solution functions are not easy to obtain analytically.

11.5.1 Computational example

We consider the second order differential equation:

$$y''(t) - 2y'(t) + y = 0.$$

This is equivalent to

$$y''(t) = 2y'(t) - y(t) = f(t, y, y').$$

The conversion is based on

$$w_1(t) = y(t),$$

and

$$w_2(t) = y'(t).$$

The resulting converted system is

$$w_1'(t) = w_2(t),$$

and
$$w_2'(t) = -w_1(t) + 2w_2(t),$$

which is the system solved in the last section. The analytic solution of

$$y(t) = e^t + te^t$$

satisfies the second order differential equation since

$$y'(t) = 2e^t + te^t,$$

and
$$y''(t) = 3e^t + te^t,$$

resulting in

$$y'' - 2y' + y = 0.$$

11.6   Linearization of second order initial value problems

Second order initial value problems are fundamental in most engineering fields, hence alternative solution techniques are of practical importance. One of those is the solution via linearization. We consider the nonhomogeneous second order problem

$$\ddot{y}(t) + a_1 \cdot \dot{y}(t) + a_0 \cdot y(t) = p(t)$$

with given initial conditions $y(t_0), \dot{y}(t_0)$ and the parameter t representing time. The lead coefficient a_2 is chosen to be unity without the loss of generality. A reformulation of this problem into a first order equation is possible by defining a vector

$$X(t) = \begin{bmatrix} \dot{y}(t) \\ y(t) \end{bmatrix}$$

and its derivative

$$\dot{X}(t) = \begin{bmatrix} \ddot{y}(t) \\ \dot{y}(t) \end{bmatrix}.$$

The right-hand side is also organized into an array

$$P(t) = \begin{bmatrix} p(t) \end{bmatrix}$$

and the solutions at a certain time t are gathered as

$$Y = \begin{bmatrix} \ddot{y}(t) \\ \dot{y}(t) \\ y(t) \end{bmatrix}.$$

With these the second order equation may be recast into two equations of

$$\dot{X} = AX + BP$$

and

$$Y = CX + EP,$$

where the new matrices are

$$A = \begin{bmatrix} -a_1 & -a_0 \\ 1 & 0 \end{bmatrix}, B = \begin{bmatrix} 1 \\ 0 \end{bmatrix},$$

and

$$C = \begin{bmatrix} -a_1 & -a_0 \\ 1 & 0 \\ 0 & 1 \end{bmatrix}, E = \begin{bmatrix} 1 \\ 0 \\ 0 \end{bmatrix}.$$

The solution of the homogeneous equation $\dot{X}(t) = AX(t)$ is

$$X_h(t) = ke^{At}.$$

The unknown integration constant with a given initial condition $X(t_0)$ becomes

$$k = e^{-At_0}X(t_0)$$

and the homogeneous solution is

$$X_h(t) = e^{A(t-t_0)}X(t_0).$$

A particular solution of the nonhomogeneous equation $\dot{X} = AX + BP$ is of the form

$$X_p(t) = k(t)e^{At}.$$

Substituting and differentiating yields the now time-dependent variable

$$k(t) = \int_{t_0}^{t} e^{-A\tau}BP(\tau)d\tau.$$

Therefore the general solution of the linearized equation becomes

$$X(t) = e^{A(t-t_0)}X(t_0) + \int_{t_0}^{t} e^{A(t-\tau)}BP(\tau)d\tau.$$

The first term is the homogeneous solution dependent only on the initial condition. The second term is the particular solution for the input load $P(t)$ in a convolution integral. The linearized solution may be efficiently evaluated by the following simple time stepping algorithm.

Loop $k = 0, 1, \ldots$

a. Compute current time $t = t_0 + k\Delta t$ with time step Δt

b. Find current load $P(t)$

c. Compute $X(t + \Delta t) = e^{Ak\Delta t} X(t_0) + \int_t^{t+\Delta t} e^{A(t+\Delta t - \tau)} BP(\tau) d\tau$

d. Update vector $X(t) = X(t + \Delta t)$

e. Extract current response $Y(t) = CX(t) + EP(t)$

f. Increment counter $k = k + 1$ and repeat from a.

The matrix exponential may be replaced by the power series approximation introduced in Section 4.6. Since the load vector is a single column, the matrix in the integral also becomes a single column vector; therefore the matrix integral becomes a term-wise scalar integral.

This method is called the state space solution in the industry and is preferred by electrical engineers and control system analysts. In practical control problems it is often desirable to make setpoint changes without overshoot and this may be accomplished by the linear state space models that are also easier to solve than quadratic problems. The approach is also applicable to mechanical engineering problems, such as the one shown in the next section.

11.7   Transient response analysis application

A prominent application of initial value problems in engineering is the calculation of the response of a mechanical system to a time-dependent excitation. The computation is called the transient response analysis of mechanical systems. The equilibrium of the system at a step t in time is

$$M\ddot{y}(t) + B\dot{y}(t) + Ky(t) = F(t).$$

Here the K and M matrices are the stiffness and mass matrices, respectively. The y is the displacement vector and $\ddot{y}$ is the acceleration. The B is the damping matrix associated with the velocity $\dot{y}$. The $F(t)$ is a time-dependent load acting on various parts of the system.

It should be mentioned that while here we focus on ordinary differential equations, initial value problems may be posed with partial differential equations as well. Some specific applications arise from modeling heat conduction and wave propagation.

The practical method often employed in the industry is the Newmark method, which is a special two-step implicit method much along the lines of those shown earlier in this chapter. Considering $h = \Delta t$, the method is based on

$$y(t + \Delta t) = \frac{1}{4}y(t + 2\Delta t) + \frac{1}{2}y(t + \Delta t) + \frac{1}{4}y(t).$$

The initial condition is that there is no acceleration when $t < 0$. It is, however, allowed that the system has a constant initial velocity defined by a nonzero value of $\dot{y}(0)$ and a starting displacement of $y(0)$. In order to start the Newmark process with an equilibrium at time $t + \Delta t = 0$, the following initial conditions are used. For the displacement,

$$y(-\Delta t) = y(0) - \dot{y}(0)\Delta t.$$

The starting load is computed as

$$F(-\Delta t) = Ky(0) + B\dot{y}(0).$$

More on the details of this method and its generalization may be found in [8].

11.7.1   Model order reduction

The above solution in practical circumstances is computationally expensive due to the large sizes of the matrices involved. There are two major ways to reduce the computational complexity: by reducing the order of the mathematical model and by minimizing the level or interaction with the physical environment.

In order to reduce the mathematical model order, we rely on the normal modes solution introduced in Section 10.8 and solve that problem for multiple eigenvalues and eigenvectors as

$$K_{ff}\Phi_{fh} = M\Phi_{fh}\Lambda_{hh}.$$

The index h indicates the number of eigenvalues and vectors obtained, Φ is the eigenvector matrix with h columns and the same number of rows (f) as the M, K matrices and Λ is an order h diagonal matrix containing the eigenvalues on the diagonal. The original transient response solution is replaced by the so-called modal solution as

$$y_f = \Phi_{fh}\gamma_h.$$

Substituting into the original problem and premultiplying it by the eigenvectors, we obtain

$$m_{hh}\ddot{\gamma}_h + b_{hh}\dot{\gamma}_h + k_{hh}\gamma_h = f_h,$$

where the new reduced order, so-called modal, matrices are

$$m_{hh} = \Phi_{fh}^T M_{ff} \Phi_{fh},$$
$$b_{hh} = \Phi_{fh}^T B_{ff} \Phi_{fh},$$
$$k_{hh} = \Phi_{fh}^T K_{ff} \Phi_{fh},$$

and the modal load vector is

$$f_h = \Phi_{fh}^T F_f.$$

The order of this modal transient response problem is now h, which is in practice several orders of magnitude smaller than the original mathematical model.

Furthermore, the reduced order matrices are of specific structure; m, k are diagonal matrices allowing a very efficient solution computation. The b matrix usually is not diagonal, however, of a smaller rank since damping is usually applied only to a limited part of the structure.

Reducing the level of interaction with the physical environment is based on a similar recognition: that excitations are also only applied at certain parts of the structure and the responses are also only evaluated at a limited set of points important to the application.

This will produce a very sparse F and y, resulting in a very efficient execution of the time steps. Let us introduce the input force mapping relation

$$F_f = \begin{bmatrix} \cdots \\ F_i \\ \cdots \\ F_j \\ \cdots \end{bmatrix} = \begin{bmatrix} \cdots\,1\,0\,0\,\cdots \\ \cdots\,0\,0\,1\,\cdots \end{bmatrix}^T \begin{bmatrix} P_1 \\ P_2 \end{bmatrix} = R_{sf}^T P_s.$$

Here the s index refers to the order of the sparse external load vector (P) and we assumed for the sake of explanation that only two points are loaded whose locations correspond to the units in the mapping matrix. Utilizing this, the modal solution right-hand side becomes

$$f_h = \Phi_{fh}^T R_{sf}^T P_s.$$

Similarly we introduce the only required output result (Y) locations as

$$Y_s = \begin{bmatrix} Y_1 \\ Y_2 \end{bmatrix} = \begin{bmatrix} \dots & 1 & 0 & 0 & \dots \\ \dots & 0 & 0 & 1 & \dots \end{bmatrix} \begin{bmatrix} \dots \\ y_k \\ \dots \\ y_l \\ \dots \end{bmatrix} = S_{sf} y_f.$$

We again assumed that there are only two points at which the results are required and their locations are indicated by the units. These locations are not necessarily the same as on the input side, hence designated by k, l as opposed to i, j. The required physical results then will be computed from the modal solution as

$$Y_s = S_{sf} \Phi_{fh} \gamma_h.$$

Finally, there are other projection approaches to reduce the order of the solution, such as

- rational polynomial interpolation using moment matching and the Padé via Lanczos technique [6], and

- proper orthogonal decomposition using a sampling technique and singular value decomposition [11].

They are not detailed here further; the reader is encouraged to review the references.

References

[1] Bajcsay, P.; *Numerikus analizis*, Tankönyvkiadó, Budapest, 1972.

[2] Euler, L; Methodus generalis summandi progressiones, *Comm. Acad. Imp. Petr.*, Vol. 6, pp. 68–97, 1738.

[3] Fehlberg, E.; Klassische Runge–Kutta Formeln vierter und niedriger Ordnung mit Schrittweiten-Kontrolle und ihre Anwendung auf Wärmeleitungsprobleme, *Computing*, Vol. 6, pp. 61–71, 1970.

[4] Gear, C. W.; *Numerical Initial Value Problems in Ordinary Differential Equations*, Prentice Hall, Englewood Cliffs, NJ, 1971.

[5] Gragg, W. B.; On extrapolation algorithms for ordinary initial value problems, *SIAM J. on Num. Anal.*, Vol. 2, pp. 384–403, 1965.

[6] Komzsik, L.; *The Lanczos Method: Evolution and Application*, SIAM, Philadelphia, 2003.

[7] Moulton, F. R.; *Differential Equations*, Dover, New York, 1958.

[8] Newmark, N. M.; A method of computation for structural dynamics, *ASME Proceedings*, 1959.

[9] Richtmyer, R. and Morton, K. W.; *Difference Methods for Initial Value Problems*, Wiley, 1967.

[10] Runge, C.; Über empirische Funktionen und die Interpolation zwischen äquidistanten ordinaten, *Zeitung für Math. Phys.*, Vol. 46, pp. 224–243, 1901.

[11] Willcox, K. and Peraire, J.; Balanced model reduction via proper orthogonal decomposition, *AIAA Journal*, Vol. 40, pp. 2323–2330, 2002.

12

Boundary value problems

Boundary value problems (BVPs) arise when some conditions are imposed on a differential equation at certain spatial points. This topic is an important subject of interest to engineers, as many physical phenomena are described by boundary value problems. Boundary value problems of various second order differential equations are used most often and will be the focus of this chapter.

The two dominant methods of solution for boundary value problems are the finite difference and the finite element methods. Each of them may be applied to both ordinary or partial differential equations based boundary value problems. In the case of multidimensional problems and irregular geometric boundaries, the finite element method has a definite advantage. In this chapter the finite difference method will be introduced in connection with ordinary differential equations, and the finite element method, with partial differential equations (PDEs) based boundary value problems.

A now classical reference on numerical solutions of boundary value problems is found in [2]. The finite difference method has its roots in the earlier difference based approaches, and an overview is found in [4]. The seminal paper anchoring the finite element method is Galerkin's from 1915 [1]. The main engineering reference for the finite element method is found in [6], and additional analysis is found in [5].

12.1 Boundary value problems of ordinary differential equations

As mentioned above, the most important problems are second order differential equations. For such, there are two conditions that may be imposed at certain locations; hence they are commonly called two-point boundary value problems.

12.1.1   Nonlinear boundary value problems

These problems in the general nonlinear case are of the form

$$y'' = f(x, y, y'); \ a \le x \le b,$$

with boundary conditions

$$y(a) = \alpha,$$

and

$$y(b) = \beta.$$

Let us assume that the function f and the partial derivatives $\frac{\partial f}{\partial y}$ and $\frac{\partial f}{\partial y'}$ are continuous on the domain

$$D = [(x, y, y') : a \le x \le b; -\infty < y < \infty; -\infty < y' < \infty].$$

Then the boundary value problem has a unique solution if

$$|\frac{\partial f}{\partial y}(x, y, y')| > 0,$$

and

$$|\frac{\partial f}{\partial y'}(x, y, y')| \le L,$$

for all $(x, y, y') \in D$ and a constant L.

12.1.2   Linear boundary value problems

The problem is simpler if the function f has a specific structure. The boundary value problem of a second order ordinary differential equation is linear if it may be brought to the form of

$$f(x, y, y') = a(x)y' + b(x)y + c(x).$$

The condition for a unique solution in this case simplifies to the following requirement:

$$b(x) > 0; x \in [a, b]$$

and $a(x), b(x)$ and $c(x)$ must be continuous on the interval $[a, b]$.

12.1.3 Approximate solution of boundary value problems

We consider the problem of Section 12.1.1 rewritten in the form of

$$y'' + p(x)y' + q(x)y = r(x),$$

with boundary conditions $y(a) = \alpha, y(b) = \beta$. Assuming there is a unique solution, we introduce two new boundary value problems:

$$u'' + p(x)u' + q(x)u = 0, u(a) = 0, u'(a) = 1,$$

and

$$z'' + p(x)z' + q(x)z = r(x), z(a) = \alpha, z'(a) = 0.$$

From the solution of these two problems the approximate solution of the original problem may be obtained as

$$y(x) = z(x) + \frac{\beta - z(b)}{u(b)} u(x).$$

Differentiating twice,

$$y'(x) = z'(x) + \frac{\beta - z(b)}{u(b)} u'(x),$$

$$y''(x) = z''(x) + \frac{\beta - z(b)}{u(b)} u''(x),$$

and substituting into the original problem, we get

$$(z''(x) + p(x)z'(x) + q(x)z(x)) + \frac{\beta - z(b)}{u(b)} (u''(x) + p(x)u'(x) + q(x)u(x)) = r(x).$$

The first term on the left is $r(x)$, the second one is zero by the new problem definition, hence we obtained an identity

$$r(x) = r(x),$$

proving that the so constructed solution satisfies the original equation.

The boundary values are also satisfied because

$$y(a) = z(a) + \frac{\beta - z(b)}{u(b)} u(a) = z(a) = \alpha,$$

and

$$y(b) = z(b) + \frac{\beta - z(b)}{u(b)} u(b) = \beta.$$

To judge whether the two replacement problems are easier to solve than the original is left to the reader.

We now consider the generic form of the problem as posed in Section 12.1.1

$$f(x, y, y', y'') = 0$$

with boundary conditions

$$y(a) = \alpha, y(b) = \beta$$

and devise an approximate solution. We select a set of linearly independent functions

$$u_0(x), u_1(x), ..., u_n(x)$$

for which the boundary conditions are

$$u_0(a) = \alpha, u_0(b) = \beta,$$

and

$$u_k(a) = u_k(b) = 0, k = 1, 2, ...n.$$

We seek the approximate solution of this problem in the form of

$$u(x) = u_0(x) + \sum_{k=1}^{n} c_k u_k(x).$$

The goal is to find the values of the c_k coefficients that produce a good approximation to the solution

$$y(x) \approx u(x).$$

There are several ways to pose this problem of finding the coefficients, such as the method of least squares or the collocation technique. The method of moments we introduce here is seeking the coefficients such that substituting the approximate solution into the f function is orthogonal to the trial functions.

$$\int_a^b f(x, u(x), u'(x), u''(x)) u_k(x) dx = 0, k = 1, 2, ..., n.$$

This method is the basis of the Galerkin method to be discussed in detail in Section 12.5.

12.2 The finite difference method for boundary value problems of ordinary differential equations

We will develop the finite difference method in connection with linear problems, although much of the discussion applies to nonlinear problems as well. The underlying principle, as the name indicates, is to employ some of the difference schemes introduced earlier in Chapter 5.

We divide the interval of interest into $n + 1$ finite segments as

$$x_i = a + ih; \quad i = 0, 1, \ldots, n + 1,$$

where

$$h = \frac{b - a}{n + 1}.$$

This subdivision is commonly called the mesh discretization, which is more meaningful in the case of a two-dimensional domain, but we will use it nevertheless. The original differential equation at these mesh points is

$$y''(x_i) = a(x_i)y'(x_i) + b(x_i)y(x_i) + c(x_i),$$

and the boundary conditions are

$$y(a) = y_0,$$

and

$$y(b) = y_{n+1}.$$

Let us approximate the as yet unknown solution function by its Taylor polynomial of degree 3 with the appropriate remainder term in the neighborhood of the point x_i.

$$y(x) = y(x_i) + (x - x_i)y'(x_i) + \frac{(x - x_i)^2}{2}y''(x_i) + \frac{(x - x_i)^3}{3!}y'''(x_i)$$
$$+ \frac{(x - x_i)^4}{4!}y^{(4)}(\xi).$$

Here, as usual, $\xi \in (x, x_i)$. Evaluating this at $x_{i+1} = x_i + h$ yields

$$y(x_{i+1}) = y(x_i) + hy'(x_i) + \frac{h^2}{2}y''(x_i) + \frac{h^3}{3!}y'''(x_i) + \frac{h^4}{4!}y^{(4)}(\xi).$$

Similarly for $x_{i-1} = x_i - h$ the equation is

$$y(x_{i-1}) = y(x_i) - hy'(x_i) + \frac{h^2}{2}y''(x_i) - \frac{h^3}{3!}y'''(x_i) + \frac{h^4}{4!}y^{(4)}(\zeta).$$

Here, $\zeta \in (x_{i-1}, x_i)$. The emerging negative signs hint at the idea of adding these equations, resulting in

$$y(x_{i+1}) + y(x_{i-1}) = 2y(x_i) + h^2 y''(x_i) + \frac{h^4}{4!}(y^{(4)}(\xi) + y^{(4)}(\zeta)).$$

Introducing

$$y^{(4)}(\xi_i) = \frac{y^{(4)}(\xi) + y^{(4)}(\zeta)}{2},$$

and solving for the second derivative, we obtain a centered difference approximation

$$y''(x_i) = \frac{1}{h^2}(y(x_{i+1}) - 2y(x_i) + y(x_{i-1})) - \frac{h^2}{12}y^{(4)}(\xi_i),$$

with $\xi_i \in (x_{i-1}, x_i)$. Similar activity with a second order Taylor polynomial results in a centered difference approximation of the first derivative (already introduced in Section 5.1.1) as

$$y'(x_i) = \frac{1}{2h}(y(x_{i+1}) - y(x_{i-1})) - \frac{h^2}{6}y^{(3)}(\zeta_i),$$

with $\zeta_i \in (x_{i-1}, x_i)$. Finally we substitute these difference formulae into the original differential equation.

$$\frac{1}{h^2}(y(x_{i+1}) - 2y(x_i) + y(x_{i-1})) = a(x_i)\frac{1}{2h}(y(x_{i+1}) - y(x_{i-1}))$$

$$+b(x_i)y(x_i) + c(x_i) - \frac{h^2}{6}(a(x_i)y^{(3)}(\zeta_i) - \frac{1}{2}y^{(4)}(\xi_i)).$$

As before, we introduce the approximate solution values y_i in place of $y(x_i)$ to obtain an approximate solution with a local error of $O(h^2)$.

$$\frac{1}{h^2}(y_{i+1} - 2y_i + y_{i-1}) = a(x_i)\frac{1}{2h}(y_{i+1} - y_{i-1}) + b(x_i)y_i + c(x_i).$$

The finite difference method is obtained by reordering in increasing mesh index order as

$$(-1 - \frac{h}{2}a(x_i))y_{i-1} + (2 + h^2 b(x_i))y_i + (-1 + \frac{h}{2}a(x_i))y_{i+1} = -h^2 c(x_i).$$

Of course we cannot lose sight of the fact that we are operating on a mesh of $i = 0, 1, \ldots, n$, hence the problem is a system of linear equations,

$$AY = B,$$

where the first three columns of A are

$$A(1:3) = \begin{bmatrix} 2+h^2b(x_1) & -1+\frac{h}{2}a(x_1) & 0 \\ -1-\frac{h}{2}a(x_2) & 2+h^2b(x_2) & -1+\frac{h}{2}a(x_2) \\ 0 & & \end{bmatrix},$$

and the last three columns are

$$A(n-2:n) = \begin{bmatrix} & \cdot & 0 \\ -1-\frac{h}{2}a(x_{n-1}) & 2+h^2b(x_{n-1}) & -1+\frac{h}{2}a(x_{n-1}) \\ 0 & -1-\frac{h}{2}a(x_n) & 2+h^2b(x_n) \end{bmatrix},$$

The solution vector is of the form

$$Y = \begin{bmatrix} y_1 \\ y_2 \\ \cdot \\ y_{n-1} \\ y_n \end{bmatrix},$$

and the right-hand side is

$$B = \begin{bmatrix} -h^2c(x_1) + (1+\frac{h}{2}a(x_1))y_0 \\ -h^2c(x_2) \\ \cdot \\ -h^2c(x_{n-1}) \\ -h^2c(x_n) + (1-\frac{h}{2}a(x_n))y_{n+1} \end{bmatrix}.$$

Note that the y_0, y_{n+1} values on the right-hand side are the boundary values, and all the other terms are known. This is an inhomogeneous linear system of equations, which has a unique solution if its determinant is not zero. For a tridiagonal matrix the determinant is nonzero if

$$|A(i, i-1)| + |A(i, i+1)| \le |A(i, i)|,$$

or the sum of the absolute values of the off-diagonal terms does not exceed the absolute value of the diagonal. Considering the special content of this matrix, this condition is met when

$$\frac{h}{2} < \frac{1}{M},$$

where

$$M = max|a(x)|; x \in [a, b].$$

There is a loosely framed message in this condition, which is that by reducing the discretization size one can always produce a solution. It is, of course, assumed that the solution function has a continuous fourth derivative, a component of the local error that assured the method's $O(h^2)$ behavior.

The solution of this tridiagonal system may be obtained efficiently, despite its asymmetry, which lessens the efficiency somewhat.

12.3   Boundary value problems of partial differential equations

For engineers the most important class of partial differential equations is that of second order partial differential equations (PDEs). Their generic form is

$$a(x,y)\frac{\partial^2 u}{\partial x^2} + 2b(x,y)\frac{\partial^2 u}{\partial x \partial y} + c(x,y)\frac{\partial^2 u}{\partial y^2}$$

$$+d(x,y)\frac{\partial u}{\partial x} + e(x,y)\frac{\partial u}{\partial y} + f(x,y)u(x,y) + g(x,y) = 0.$$

They are commonly classified as hyperbolic when

$$b^2 - a \cdot c > 0,$$

parabolic when

$$b^2 - a \cdot c = 0,$$

and elliptic when

$$b^2 - a \cdot c < 0.$$

Note that since the coefficients are functions, the differential equations may have different types in different domains. For example, the partial differential equation

$$\frac{\partial^2 u}{\partial x^2} + (1 - x^2 - y^2)\frac{\partial^2 u}{\partial y^2} = 0$$

has $a = 1, b = 0$ and $c(x,y) = 1 - x^2 - y^2$, hence

$$b^2 - a \cdot c = x^2 + y^2 - 1.$$

This quantity is negative when the (x,y) point is inside the unit circle and positive outside. Therefore, this differential equation is elliptic inside the unit circle and hyperbolic outside the circle. The unit circle represents the boundary condition when solving either area.

A boundary value problem in connection with a partial differential equation is posed as an equation of the above generic class, defined or interpreted on a domain D and with some conditions prescribed on the whole or on a part of the boundary B of the domain. If the boundary conditions are prescribed as

$$u(x, y) = h(x, y); \ (x, y) \in B,$$

they are called Dirichlet boundary conditions.

A partial differential equation of considerable interest to engineers is the second order elliptic partial differential (Laplace's) equation:

$$\Delta u(x, y) = \frac{\partial^2 u}{\partial x^2} + \frac{\partial^2 u}{\partial y^2} = 0.$$

This equation appears in the modeling of several physical phenomena, such as the steady-state distribution of heat in a planar domain.

We will also consider its variant, Poisson's equation, which is stated as

$$\Delta u(x, y) - \frac{\partial^2 u}{\partial x^2} + \frac{\partial^2 u}{\partial y^2} = -f(x, y).$$

The most commonly known physical phenomenon is the deformation of a planar, clamped membrane under a pressure load. Both of these equations are used in the demonstration of the theory and in computational examples.

12.4 The finite difference method for boundary value problems of partial differential equations

The finite difference method is also applicable to partial differential equations (PDEs). We will consider Laplace's equation on the rectangular, two-dimensional domain

$$D = [(x, y) : a_x \le x \le b_x; a_y \le y \le b_y].$$

The boundary conditions applied at the B perimeter of this domain,

$$u(x, y) = h(x, y); (x, y) \in B.$$

As this equation is now a function of two spatial variables, we discretize in both directions. There is no need to choose the same step sizes. We use

$$x_i = a_x + ih,$$

and

$$y_i = a_y + jk,$$

with

$$h = \frac{b_x - a_x}{n + 1},$$

and

$$k = \frac{b_y - a_y}{m + 1}.$$

The constant values of

$$x = x_i; y = y_j$$

are called grid lines and now they truly constitute a mesh. The Taylor series expression in Section 12.2 is easy to extend along a certain horizontal mesh line y_j as

$$\frac{\partial^2 u}{\partial x^2}(x_i, y_j) = \frac{1}{h^2}(u(x_{i+1}, y_j) - 2u(x_i, y_j) + u(x_{i-1}, y_j)) - \frac{h^2}{12}\frac{\partial^4 u}{\partial x^4}(\xi_i, y_j),$$

where $\xi_i \in (x_{i-1}, x_{i+1})$ as before. The extension along a vertical mesh line x_i is also straightforward:

$$\frac{\partial^2 u}{\partial y^2}(x_i, y_j) = \frac{1}{k^2}(u(x_i, y_{j+1}) - 2u(x_i, y_j) + u(x_i, y_{j-1})) - \frac{k^2}{12}\frac{\partial^4 u}{\partial y^4}(x_i, \zeta_j),$$

with $\zeta_j \in (y_{i-1}, y_{i+1})$. With these difference formulae, Laplace's equation, the current subject of our discussion, becomes

$$\frac{1}{h^2}(u(x_{i+1}, y_j) - 2u(x_i, y_j) + u(x_{i-1}, y_j)) +$$

$$\frac{1}{k^2}(u(x_i, y_{j+1}) - 2u(x_i, y_j) + u(x_i, y_{j-1})) = \frac{h^2}{12}\frac{\partial^4 u}{\partial x^4}(\xi_i, y_j) + \frac{k^2}{12}\frac{\partial^4 u}{\partial y^4}(x_i, \zeta_j).$$

Here $i = 1, 2, \ldots, n$ and $j = 1, 2, \ldots, m$. The $i = 0, n + 1$ and $j = 0, m + 1$ index combinations represent the boundary conditions as

$$u(x_0, y_j) = h(x_0, y_j) = u_{0,j},$$

and

$$u(x_{n+1}, y_j) = h(x_{n+1}, y_j) = u_{n+1,j},$$

for $j = 0, 1, \ldots, m + 1$. Similarly for $i = 0, 1, \ldots, n + 1$,

$$u(x_i, y_0) = h(x_i, y_0) = u_{i,0},$$

and
$$u(x_i, y_{m+1}) = h(x_i, y_{m+1}) = u_{i,m+1}.$$

Finally introducing the approximate values $u_{i,j}$, and reordering by increasing indices, results in the finite difference method for this problem.

$$-u_{i-1,j} + (2\frac{h^2}{k^2} + 2)u_{i,j} - u_{i+1,j} - \frac{h^2}{k^2}(u_{i,j+1} + u_{i,j-1}) = 0.$$

The equation shows a very distinct "cross" pattern: in each center grid location of (i, j), its vertical $(i, j+1), (i, j-1)$ and horizontal $(i-1, j), (i+1, j)$ neighbors are used. For the specific case of a square region R and the same step size in both directions, the equation simplifies to

$$-u_{i-1,j} + 4u_{i,j} - u_{i+1,j} - u_{i,j+1} - u_{i,j-1} = 0.$$

Remember, of course, that $i = 1, 2, \ldots, n$ and $j = 1, 2, \ldots, m$. For an orderly solution of the whole interior domain via a linear system, we introduce a matrix A and a vector U. Their parts corresponding to the (i, j) equation are

$$A = \begin{bmatrix} . & . & . & . & . & . & . & . & . & . & . \\ . & 4 & -1 & 0 & -1 & 0 & 0 & 0 & 0 & 0 & . \\ . & -1 & 4 & -1 & 0 & -1 & 0 & 0 & 0 & 0 & . \\ . & 0 & -1 & 4 & 0 & 0 & -1 & 0 & 0 & 0 & . \\ . & -1 & 0 & 0 & 4 & -1 & 0 & -1 & 0 & 0 & . \\ . & 0 & -1 & 0 & -1 & 4 & -1 & 0 & -1 & 0 & . \\ . & 0 & 0 & -1 & 0 & -1 & 4 & 0 & 0 & -1 & . \\ . & 0 & 0 & 0 & -1 & 0 & 0 & 4 & -1 & 0 & . \\ . & 0 & 0 & 0 & 0 & -1 & 0 & -1 & 4 & -1 & . \\ . & 0 & 0 & 0 & 0 & 0 & -1 & 0 & -1 & 4 & . \\ . & . & . & . & . & . & . & . & . & . & . \end{bmatrix},$$

and

$$U = \begin{bmatrix} . \\ u_{i-1,j+1} \\ u_{i,j+1} \\ u_{i+1,j+1} \\ u_{i-1,j} \\ u_{i,j} \\ u_{i+1,j} \\ u_{i-1,j-1} \\ u_{i,j-1} \\ u_{i+1,j-1} \\ . \end{bmatrix}.$$

Note that here the diagonal neighbors of (i, j) are also included. The solution comes from the linear system

$$AU = B.$$

Note, however, that despite the fact that there was no right-hand side of the differential equation, some of the terms in the B vector are not zero, they represent the boundary conditions. For example, in the ordering chosen above,

$$B(1) = u_{0,m+1},$$

$$B((m+1)(n+1)) = u_{n+1,0},$$

and so on. Hence, the method is also applicable to the same differential equation with a nonzero right-hand side (Poisson's equation), which will be discussed in more detail in the upcoming sections.

12.4.1 Computational example

The example we choose to demonstrate this technique is Laplace's equation:

$$\Delta u(x,y) = \frac{\partial^2 u}{\partial x^2} + \frac{\partial^2 u}{\partial y^2} = 0,$$

on the domain of the unit square located in the first quadrant

$$D = [(x,y); 0 < x < 1; 0 < y < 1],$$

and

$$B = [x = 0; y = 0; x = 1; y = 1].$$

The boundary conditions are described as

$$u(0,y) = 0; u(x,0) = 0,$$

and

$$u(x,1) = x; u(1,y) = y.$$

Some observation shows that this corresponds to the analytic solution

$$u(x,y) = xy,$$

since

$$\frac{\partial^2 u}{\partial x^2} + \frac{\partial^2 u}{\partial y^2} = 0 + 0 = 0.$$

The solution surface is shown in Figure 12.1.

We will use a uniform step size of

$$h = 1/2, k = 1/2,$$

resulting in $n = 1$ and $m = 1$ as well as in a 3 by 3 finite difference mesh with

$$x_0 = 0, x_1 = 1/2, x_2 = 1,$$

and

$$y_0 = 0, y_1 = 1/2, y_2 = 1.$$

For this case $i = 0, 1, 2$ and $j = 0, 1, 2$, but the locations with $i = 0, 2$ and $j = 0, 2$ belong to the boundary. Hence there is only one interior point of the solution, resulting in one equation:

$$-u_{0,1} + 4u_{1,1} - u_{2,1} - u_{1,2} - u_{1,0} = 0.$$

Substituting the boundary values above, this equation becomes

$$-0 + 4u_{1,1} - 1/2 - 1/2 - 0 = 0,$$

from which the approximate solution is

$$u_{1,1} = 1/4.$$

The exact solution at this location $(1/2, 1/2)$ is

$$u(1/2, 1/2) = 1/4,$$

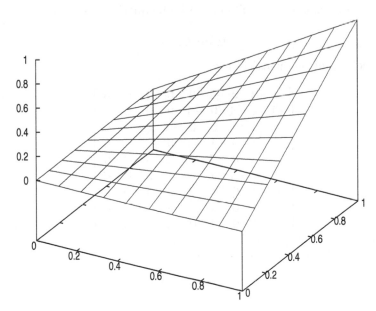

FIGURE 12.1 Finite difference example solution

which is the same. This is not a surprise, since the local error is proportional to the fourth derivative of the function, which is zero.

TABLE 12.1
Finite difference example

x_i	0	1/4	1/2	3/4	1
j,i	0	1	2	3	4
4	0	1/4	1/2	3/4	1
3	0	$u_{1,3}$	$u_{2,3}$	$u_{3,3}$	3/4
2	0	$u_{1,2}$	$u_{2,2}$	$u_{3,2}$	1/2
1	0	$u_{1,1}$	$u_{2,1}$	$u_{3,1}$	1/4
0	0	0	0	0	0

Let us now consider a finer discretization for demonstration purposes. Assume $n = 3, m = 3$ or $h = 1/4, k = 1/4$. The finite difference mesh now would become a 5 by 5 mesh and 9 interior points would be computed. The assignments are shown in Table 12.1 which shows the yet unknown interior approximate solutions, and the surrounding constants are the boundary values.

This finer approximation results in a 9 by 9 equation system of

$$AU = B,$$

where the left-hand side matrices are

$$A = \begin{bmatrix} 4 & -1 & 0 & -1 & 0 & 0 & 0 & 0 & 0 \\ -1 & 4 & -1 & 0 & -1 & 0 & 0 & 0 & 0 \\ 0 & -1 & 4 & 0 & 0 & -1 & 0 & 0 & 0 \\ -1 & 0 & 0 & 4 & -1 & 0 & -1 & 0 & 0 \\ 0 & -1 & 0 & -1 & 4 & -1 & 0 & -1 & 0 \\ 0 & 0 & -1 & 0 & -1 & 4 & 0 & 0 & -1 \\ 0 & 0 & 0 & -1 & 0 & 0 & 4 & -1 & 0 \\ 0 & 0 & 0 & 0 & -1 & 0 & -1 & 4 & -1 \\ 0 & 0 & 0 & 0 & 0 & -1 & 0 & -1 & 4 \end{bmatrix},$$

and

$$U = \begin{bmatrix} u_{1,3} \\ u_{2,3} \\ u_{3,3} \\ u_{1,2} \\ u_{2,2} \\ u_{3,2} \\ u_{1,1} \\ u_{2,1} \\ u_{3,1} \end{bmatrix}.$$

The right-hand side is now constructed with the given boundary conditions as

$$B = \begin{bmatrix} u_{0,3} + u_{1,4} \\ u_{2,4} \\ u_{4,3} + u_{3,4} \\ u_{0,2} \\ 0 \\ u_{4,2} \\ u_{0,1} + u_{1,0} \\ u_{2,0} \\ u_{3,0} + u_{4,1} \end{bmatrix}.$$

This topic was mentioned conceptually in the last section, but now we are in the position to clarify it. Substituting the actual values yields

$$B = \begin{bmatrix} 0 + 1/4 \\ 1/2 \\ 3/4 + 3/4 \\ 0 \\ 0 \\ 1/2 \\ 0 + 0 \\ 0 \\ 0 + 1/4 \end{bmatrix} = \frac{1}{16} \begin{bmatrix} 4 \\ 8 \\ 24 \\ 0 \\ 0 \\ 8 \\ 0 \\ 0 \\ 4 \end{bmatrix}.$$

The solution of this system is

$$U = \begin{bmatrix} 3/16 \\ 3/8 \\ 9/16 \\ 1/8 \\ 1/4 \\ 3/8 \\ 1/16 \\ 1/8 \\ 3/16 \end{bmatrix} = \frac{1}{16} \begin{bmatrix} 3 \\ 6 \\ 9 \\ 2 \\ 4 \\ 6 \\ 1 \\ 2 \\ 3 \end{bmatrix}.$$

This, of course, is also the exact solution, for the same reason the coarse, single equation solution was. We now introduce a different and also very practical approach to solve boundary value problems of partial differential equations, the famed finite element method.

12.5   The finite element method

The boundary value problem we now attempt to solve approximately is Poisson's equation

$$-\frac{\partial^2 u}{\partial x^2} - \frac{\partial^2 u}{\partial y^2} = f(x,y),$$

where $u(x,y)$ is the solution value at (x,y) and $f(x,y)$ is the function acting on the domain. The boundary conditions are given on the perimeter B of the now general and not necessarily rectangular domain D. For the simplicity of this discussion we will assume zero boundary conditions,

$$u(x,y) = 0; (x,y) \in B,$$

but that is not required.

We approach this approximation problem by using the method of weighted residuals, which requires that the residual of the following integral form is zero.

$$\int\int(-\frac{\partial^2 u}{\partial x^2} - \frac{\partial^2 u}{\partial y^2})w(x,y)dx\,dy - \int\int f(x,y)w(x,y)dx\,dy = 0.$$

Here and in the following, the double integrals are taken over the two-dimensional domain D of the boundary value problem and the integral boundary is omitted for simplicity's sake. The $w(x,y)$ is a weighting function. Integrating the first integral by parts and moving the second one to the right-hand side yields the so-called weak form of the boundary value problem:

$$\int\int(\frac{\partial u}{\partial x}\frac{\partial w}{\partial x} + \frac{\partial u}{\partial y}\frac{\partial w}{\partial y})dx\,dy = \int\int f(x,y)w(x,y)dx\,dy.$$

Now we apply Galerkin's method [1]. The idea is to approximate the solution as

$$u(x,y) = u_1 N_1 + u_2 N_2 + ... + u_n N_n,$$

where the u_i are the solution values at some discrete locations and

$$N_i, \quad i = 1, \ldots, n$$

is the set of finite element shape functions to be discussed in detail later in this chapter. The second, crucial part of Galerkin's method is to use the same shape functions as the weight functions:

$$w(x, y) = N_i(x, y).$$

The discrete locations are the node points of the finite element mesh. The finite element mesh is different from the finite difference mesh. The finite element mesh is based on a repeated application of triangles to cover the planar domain, as shown in Figure 12.2. In practical applications, of course, the domains are irregular and the triangles are general as well as different in size and shape.

Other objects, like rectangles or trapezoidal shapes, may also be used. The process is called meshing. The points inside the domain and on the boundary are the node points. They define the finite element mesh.

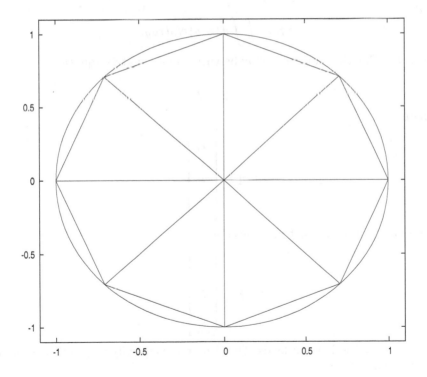

FIGURE 12.2 Finite element meshing

There may be small gaps between the boundary and the sides of the triangles adjacent to the boundary. This discrepancy contributes to the approximation error of the finite element method. It is an even bigger problem in the
finite difference method. At least in the case of the finite element method, the
gaps may be filled with progressively smaller elements, or those triangles may
be replaced with triangles having curved edges.

Substituting into the weak form of the problem, we obtain Galerkin's equations:

$$\int \int ((u_1 \frac{\partial N_1}{\partial x} + ... + u_n \frac{\partial N_n}{\partial x}) \frac{\partial N_i}{\partial x} + (u_1 \frac{\partial N_1}{\partial y} + ... + u_n \frac{\partial N_n}{\partial y}) \frac{\partial N_i}{\partial y}) dx dy =$$

$$\int \int f(x,y) N_i dx dy$$

for $i = 1, .., n$. We introduce the notation for $j = 1, 2, \ldots, n$,

$$A_{ij} = A_{ji} = \int \int (\frac{\partial N_i}{\partial x} \frac{\partial N_j}{\partial x} + \frac{\partial N_i}{\partial y} \frac{\partial N_j}{\partial y}) dx dy$$

and

$$f_i = \int \int f(x,y) N_i dx dy.$$

Then the Galerkin equations may be written as a matrix equation:

$$A\underline{u} = \underline{f},$$

where

$$\underline{u} = \begin{bmatrix} u_1 \\ . \\ u_i \\ . \\ u_n \end{bmatrix}$$

is the array of node point solutions and

$$\underline{f} = \begin{bmatrix} f_1 \\ . \\ f_i \\ . \\ f_n \end{bmatrix}$$

is the array of node point loads. The A matrix is usually very sparse as many
A_{ij} become zero.

12.5.1   Finite element shape functions

To approximate the solution inside the domain of the boundary value problem, we use piecewise polynomials. For a triangular discretization of a two-dimensional domain, bilinear interpolation functions of the form

$$u(x, y) = a + bx + cy$$

are commonly used. In order to find their coefficients, let us consider the triangular region (element) of the $x - y$ plane with corner nodes $(x_1, y_1), (x_2, y_2)$ and (x_3, y_3). For triangles, we seek shape functions satisfying

$$N_1 + N_2 + N_3 = 1.$$

Specifically, we also require that the nonzero shape function at a certain node point reduce to zero at the other two nodes, respectively. The interpolations are continuous across the neighboring elements. On an edge between two triangles, the approximation is linear and the same when approached from either element. Specifically, along the edge between nodes 1 and 2, the shape function N_3 is zero. The shape functions N_1 and N_2 along this edge are the same when calculated from an element on either side of that edge.

The solution for all the nodes of a triangular element e can be expressed as

$$u_e = \begin{bmatrix} u_1 \\ u_2 \\ u_3 \end{bmatrix} = \begin{bmatrix} 1 & x_1 & y_1 \\ 1 & x_2 & y_2 \\ 1 & x_3 & y_3 \end{bmatrix} \begin{bmatrix} a \\ b \\ c \end{bmatrix}.$$

This system of equations is solved for the unknown coefficients as

$$\begin{bmatrix} a \\ b \\ c \end{bmatrix} = \begin{bmatrix} N_{1,1} & N_{1,2} & N_{1,3} \\ N_{2,1} & N_{2,2} & N_{2,3} \\ N_{3,1} & N_{3,2} & N_{3,3} \end{bmatrix} \begin{bmatrix} u_1 \\ u_2 \\ u_3 \end{bmatrix}.$$

By substituting into the matrix form of the bilinear interpolation function

$$u(x, y) = \begin{bmatrix} 1 & x & y \end{bmatrix} \begin{bmatrix} a \\ b \\ c \end{bmatrix} = \begin{bmatrix} 1 & x & y \end{bmatrix} \begin{bmatrix} N_{1,1} & N_{1,2} & N_{1,3} \\ N_{2,1} & N_{2,2} & N_{2,3} \\ N_{3,1} & N_{3,2} & N_{3,3} \end{bmatrix} \begin{bmatrix} u_1 \\ u_2 \\ u_3 \end{bmatrix},$$

we get

$$u(x, y) = \begin{bmatrix} N_1 & N_2 & N_3 \end{bmatrix} \begin{bmatrix} u_1 \\ u_2 \\ u_3 \end{bmatrix}.$$

Here the N_1, N_2, N_3 are shape functions. Their values are

$$N_1 = N_{1,1} + N_{2,1}x + N_{3,1}y,$$

$$N_2 = N_{1,2} + N_{2,2}x + N_{3,2}y,$$

and

$$N_3 = N_{1,3} + N_{2,3}x + N_{3,3}y.$$

The shape functions, as their name indicates, clearly depend on the coordinates of the corner nodes, hence the shape of the particular triangular element of the domain. With these we are now able to approximate the relationship between the solution value inside an element in terms of the solutions at the corner node points,

$$u(x, y) = N_1 u_1 + N_2 u_2 + N_3 u_3.$$

12.5.2   Finite element matrix generation and assembly

In order to compute the $A_{i,j}$ terms of the matrix A, we proceed element by element. We consider all the nodes bounding a particular element and compute all the partial $\overline{A}_{i,j}$ terms produced by that particular element. Thus,

$$A_e = \sum_{(i,j) \in e} \overline{A}_{i,j}.$$

Finally, the A matrix is assembled as

$$A = \sum_{e=1}^{m} A_e,$$

where the summation is based on the topological relation of elements. If our element, for example, is the element described by nodes 1, 2 and 3, then the terms in A_e contribute to the terms of the first, second and third columns and rows of the global A matrix.

Let us assume that another element is adjacent to the edge between nodes 2 and 3, its other node being 4. Then by similar arguments, the second element's matrix terms (depending on that particular element's shape) will contribute to the second, third and fourth columns and rows of the global matrix. This process is continued for all the elements contained in the finite element discretization.

Then we compute the derivatives of the shape functions as follows:

$$\begin{bmatrix} \frac{\partial u(x,y)}{\partial x} \\ \frac{\partial u(x,y)}{\partial y} \end{bmatrix} = \begin{bmatrix} \frac{\partial N_1}{\partial x} & \frac{\partial N_2}{\partial x} & \frac{\partial N_3}{\partial x} \\ \frac{\partial N_1}{\partial y} & \frac{\partial N_2}{\partial y} & \frac{\partial N_3}{\partial y} \end{bmatrix} \begin{bmatrix} u_1 \\ u_2 \\ u_3 \end{bmatrix} = \begin{bmatrix} B_{1,1} & B_{1,2} & B_{1,3} \\ B_{2,1} & B_{2,2} & B_{2,3} \end{bmatrix} \begin{bmatrix} u_1 \\ u_2 \\ u_3 \end{bmatrix} = Bu_e.$$

With the above, the A_e matrix related to one finite element is

$$A_e = \int \int B^T B dx dy.$$

The resulting element matrix is of order 3 by 3 and contains all the appropriate shape function derivative combinations. The entries of A_e depend only on the shape of the element.

In practical implementations of the finite element method, the element matrices are generated as a transformation from a standard, parametrically defined element. For these cases the shape functions and their derivatives may be precomputed and appropriately transformed. The integrals posed above are usually evaluated in the industry with Gaussian quadrature, a method that was discussed in Section 6.3.

Quadrilateral planar elements are also available and used in practice. Furthermore, for more complex physical phenomena, three-dimensional tetrahedral and hexahedral elements are also used. The applicability and generality of the method is without peer. The following computational example is chosen to demonstrate the technique.

12.5.3   Computational example

We use Poisson's equation:

$$-\frac{\partial^2 u}{\partial x^2} - \frac{\partial^2 u}{\partial y^2} = f(x,y),$$

on the domain of the unit square again,

$$D = [(x,y); 0 < x < 1; 0 < y < 1],$$

with zero valued boundary conditions and the right-hand side load of

$$f(x,y) = 2\pi^2 sin(\pi x)sin(\pi y).$$

The analytic solution may easily be deduced as

$$u(x,y) = sin(\pi x)sin(\pi y).$$

That will be used to evaluate the numerical solution we will pursue. The problem and solution are shown in Figure 12.3. Since

$$\frac{\partial u}{\partial x} = \pi cos(\pi x)sin(\pi y),$$

then

$$-\frac{\partial^2 u}{\partial x^2} = \pi^2 sin(\pi x)sin(\pi x).$$

The other partial derivative behaves similarly; therefore

$$-\frac{\partial^2 u}{\partial x^2} - \frac{\partial^2 u}{\partial y^2} = 2\pi^2 sin(\pi x)sin(\pi y) = f(x,y).$$

For the feasibility of hand calculation, we will use the simple four-element finite element discretization shown in Figure 12.4 achieved by two single lines $(I1(x), I2(x))$ connecting the opposite corners of the rectangular domain. We introduced one finite element node on each of the corner points of the boundary and one in the interior, albeit having more in the interior than on the boundary is the norm in finite element meshing. The nodes are shown in Table 12.2. The first row shows the node indices and the second and third rows show the coordinates. These nodes define four triangular finite elements, one bounded by nodes 1, 2 and 3, another one by nodes 2, 3 and 4, the next one by nodes 3, 4, 5 and the last one by 1, 3, 5.

We can exploit the congruence of these four triangles by computing only one element matrix and transforming it to derive the other three. We have

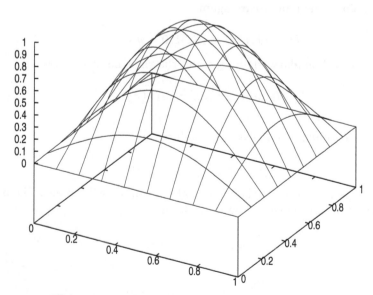

FIGURE 12.3 Poisson's equation example solution

TABLE 12.2
Finite element
example nodes

i	1	2	3	4	5
x_i	0	1	1/2	1	0
y_i	0	0	1/2	1	1

also alluded to the practical implementation concept of using a parametric standard triangle. We will just use the steps laid out in the last section for the first element for the demonstration.

Element 1.

Computation of element shape functions:

FIGURE 12.4 Finite element example mesh

$$\overline{N}_1 = \begin{bmatrix} N_{1,1} & N_{1,2} & N_{1,3} \\ N_{2,1} & N_{2,2} & N_{2,3} \\ N_{3,1} & N_{3,2} & N_{3,3} \end{bmatrix} = \begin{bmatrix} 1 & x_1 & y_1 \\ 1 & x_2 & y_2 \\ 1 & x_3 & y_3 \end{bmatrix}^{-1} = \begin{bmatrix} 1 & 0 & 0 \\ 1 & 1 & 0 \\ 1 & 1/2 & 1/2 \end{bmatrix}^{-1} = \begin{bmatrix} 1 & 0 & 0 \\ -1 & 1 & 0 \\ -1 & -1 & 2 \end{bmatrix}.$$

The shape functions are

$$N_1 = 1 - x - y,$$

$$N_2 = x - y,$$

and

$$N_3 = 2y.$$

Computation of the shape function derivatives yields

$$\begin{bmatrix} B_{1,1} & B_{1,2} & B_{1,3} \\ B_{2,1} & B_{2,2} & B_{3,3} \end{bmatrix} = \begin{bmatrix} \frac{\partial N_1}{\partial x} & \frac{\partial N_2}{\partial x} & \frac{\partial N_3}{\partial x} \\ \frac{\partial N_1}{\partial y} & \frac{\partial N_2}{\partial y} & \frac{\partial N_3}{\partial y} \end{bmatrix} = \begin{bmatrix} -1 & 1 & 0 \\ -1 & -1 & 2 \end{bmatrix}$$

and an element matrix integrand of

$$B^T B = \begin{bmatrix} 2 & 0 & -2 \\ 0 & 2 & -2 \\ -2 & -2 & 4 \end{bmatrix}.$$

The integral needs to be in two parts:

$$A_1 = \int_{x=0}^{1/2} \int_{y=0}^{x} B^T B \, dy dx + \int_{x=1/2}^{1} \int_{y=0}^{1-x} B^T B \, dy dx.$$

Since $B^T B$ is constant for our case (not always true!), the integrals are simple.

$$\int_{x=0}^{1/2} \int_{y=0}^{x} dy dx = \int_{x=0}^{1/2} x dx = [\frac{x^2}{2}]_0^{1/2} = 1/8.$$

Similarly,

$$\int_{x=1/2}^{1} \int_{y=0}^{1-x} dy dx = \int_{x=1/2}^{1} (1-x) dx = [x - \frac{x^2}{2}]_{1/2}^{1} = 1/8.$$

Their sum is, of course, the area of the element: 1/4, as may be computed by easier means. The element matrix for element 1 is then

$$A_1 = \frac{1}{4} \begin{bmatrix} 2 & 0 & -2 \\ 0 & 2 & -2 \\ -2 & -2 & 4 \end{bmatrix}.$$

Similar computations, based on nodes $2, 3$ and 4, produce the element matrix for element 2 as

$$\overline{N}_2 = \begin{bmatrix} 0 & 2 & -1 \\ 1 & -2 & 1 \\ -1 & 0 & 1 \end{bmatrix},$$

$$B_2 = \begin{bmatrix} 1 & -2 & 1 \\ -1 & 0 & 1 \end{bmatrix},$$

$$A_2 = \frac{1}{4}\begin{bmatrix} 2 & -2 & 0 \\ -2 & 4 & -2 \\ 0 & -2 & 2 \end{bmatrix}.$$

The element matrix for element 3 defined by nodes $3, 4$ and 5 is computed as

$$\overline{N}_3 = \begin{bmatrix} 2 & -1 & 0 \\ 0 & 1 & -1 \\ -2 & 1 & 1 \end{bmatrix},$$

$$B_3 = \begin{bmatrix} 0 & 1 & -1 \\ -2 & 1 & 1 \end{bmatrix},$$

$$A_3 = \frac{1}{4}\begin{bmatrix} 4 & -2 & -2 \\ -2 & 2 & 0 \\ -2 & 0 & 2 \end{bmatrix}.$$

Finally the element matrix for element 4 of nodes $1, 3, 5$ is

$$\overline{N}_4 = \begin{bmatrix} 1 & 0 & 0 \\ -1 & 2 & -1 \\ -1 & 0 & 1 \end{bmatrix},$$

$$B_4 = \begin{bmatrix} -1 & 2 & -1 \\ -1 & 0 & 1 \end{bmatrix},$$

$$A_4 = \frac{1}{4}\begin{bmatrix} 2 & -2 & 0 \\ -2 & 4 & -2 \\ 0 & -2 & 2 \end{bmatrix}.$$

Assembly of element matrices:

The individual element matrices are mapped to the global matrix that is of size 5 by 5, reflecting the presence of the 5 nodes.

$$A_1^g = \begin{bmatrix} A_1(1,1) & A_1(1,2) & A_1(1,3) & 0 & 0 \\ A_1(2,1) & A_1(2,2) & A_1(2,3) & 0 & 0 \\ A_1(3,1) & A_1(3,2) & A_1(3,3) & 0 & 0 \\ 0 & 0 & 0 & 0 & 0 \\ 0 & 0 & 0 & 0 & 0 \end{bmatrix}.$$

Similarly, considering the node locations defining the elements,

$$A_2^g = \begin{bmatrix} 0 & 0 & 0 & 0 & 0 \\ 0 & A_2(1,1) & A_2(1,2) & A_2(1,3) & 0 \\ 0 & A_2(2,1) & A_2(2,2) & A_2(2,3) & 0 \\ 0 & A_2(3,1) & A_2(3,2) & A_2(3,3) & 0 \\ 0 & 0 & 0 & 0 & 0 \end{bmatrix},$$

and

$$A_3^g = \begin{bmatrix} 0 & 0 & 0 & 0 & 0 \\ 0 & 0 & 0 & 0 & 0 \\ 0 & 0 & A_3(1,1) & A_3(1,2) & A_3(1,3) \\ 0 & 0 & A_3(2,1) & A_3(2,2) & A_3(2,3) \\ 0 & 0 & A_3(3,1) & A_3(3,2) & A_3(3,3) \end{bmatrix}.$$

The final element is somewhat different as its columns are not adjacent in the global matrix.

$$A_4^g = \begin{bmatrix} A_4(1,1) & 0 & A_4(1,2) & 0 & A_4(1,3) \\ 0 & 0 & 0 & 0 & 0 \\ A_4(2,1) & 0 & A_4(2,2) & 0 & A_4(2,3) \\ 0 & 0 & 0 & 0 & 0 \\ A_4(3,1) & 0 & A_4(3,2) & 0 & A_4(3,3) \end{bmatrix}.$$

The assembled finite element matrix is simply

$$A = \sum_1^4 A_i^g = \frac{1}{4} \begin{bmatrix} 4 & 0 & -4 & 0 & 0 \\ 0 & 4 & -4 & 0 & 0 \\ -4 & -4 & 16 & -4 & -4 \\ 0 & 0 & -4 & 4 & 0 \\ 0 & 0 & -4 & 0 & 4 \end{bmatrix} = \begin{bmatrix} 1 & 0 & -1 & 0 & 0 \\ 0 & 1 & -1 & 0 & 0 \\ -1 & -1 & 4 & -1 & -1 \\ 0 & 0 & -1 & 1 & 0 \\ 0 & 0 & -1 & 0 & 1 \end{bmatrix}.$$

From the boundary conditions it follows that all but one of the solution components are constrained.

$$\underline{u} = \begin{bmatrix} 0 \\ 0 \\ \underline{u}(3) \\ 0 \\ 0 \end{bmatrix}.$$

Note that unlike the finite difference solution, the boundary is part of the finite element solution and in this case it is of course zero, due to the imposed boundary conditions. The only free solution component (unburdened by boundary constraint) is the solution at node 3 in the middle of the domain. The rows and columns corresponding to the constraints of the yet singular A matrix are removed, resulting in the final equation of

$$A(3,3)\underline{u}(3) = \underline{f}(3).$$

The right-hand side term is computed as

$$\underline{f}(3) = \sum_{e=1}^{4} f_e^3,$$

where f_e^3 indicates the force contribution of an element to the third global node. This step is rather complicated as all four elements of the model are connected to the global node 3. We need to select the shape functions from each element that are nonzero at global node 3.

The shape function related to global node 3 of element 1 is

$$N_3 = 2y.$$

Similarly, the shape function related to global node 3 of element 2 is

$$N_2 = 2 - 2x;$$

for element 3 it is

$$N_1 = 2 - 2y,$$

and for the last element,

$$N_2 = 2x.$$

For the very first element this computation proceeds as

$$f_1^3 = 2\pi^2 \Big(\int_{x=0}^{1/2} sin(\pi x)2 \int_{y=0}^{x} ysin(\pi y)dydx$$

$$+ \int_{x=1/2}^{1} sin(\pi x)2 \int_{y=0}^{1-x} ysin(\pi y)dydx \Big).$$

As this element is symmetric, the two parts of this integral have the same value; therefore, it is enough to compute one half. Hence

$$f_1^3/2 = 2\pi^2 \frac{2}{\pi^2} \Big(\frac{1}{4} - \frac{1}{8} \Big) = \frac{1}{2}.$$

The result of the complete element integral is therefore

$$f_1^3 = 1.$$

The symmetry considerations extend cyclically to the other elements, so as a shortcut we will just apply the same value for the other contributions.

$$\underline{f}(3) = 1 + 1 + 1 + 1 = 4.$$

This would yield the final solution of

$$\underline{u}(3) = 4/4 = 1,$$

which is also the analytic solution as we executed the integration steps of the load computation exactly. The load computation is much easier when the load function is not a distributed function as in our case, but a discrete (node based) load.

Based on this nodal solution of $u_3 = 1$ we can evaluate some locations inside the elements based on

$$u(x,y) = N_1(x,y)u_1 + N_2(x,y)u_2 + N_3(x,y)u_3.$$

Inside the first element, with the use of its shape functions, one uses

$$u(x,y) = (1 - x - y)u_1 + (x - y)u_2 + 2yu_3.$$

For example, the point $(x,y) = (1/2, 1/4)$ is inside the first element and the solution is computed as

$$u(1/2, 1/4) = 1/4 \cdot 0 + 1/4 \cdot 0 + 2\frac{1}{4}1 = \frac{1}{2}.$$

The exact solution of that point is

$$u(1/2, 1/4) = sin(\frac{\pi}{2})sin(\frac{\pi}{4}) = \frac{1}{\sqrt{2}}.$$

The difference inside the element is now noticeable and due to the linear approximation within.

12.6 Finite element analysis of three-dimensional continuum

This section aims to demonstrate the finite element modeling of a three-dimensional continuum and to address some of its practical considerations, such as the use of local coordinate systems for elements and the use of the Gaussian quadrature for integration.

12.6.1 Tetrahedral finite element

We introduce a tetrahedron element, such as the one shown in Figure 12.5, located generally in global coordinates and represented also by a special local, oblique, parametric coordinate system. This is the most common object resulting from the discretization of a three-dimensional continuum.

The u,v and w are the parametric coordinate directions shown by the arrows in Figure 12.5. The u axis points from node 1 to 2, the v axis from node 1 to 3 and the w axis from node 1 to 4. This is an arbitrary selection and any other order may also be chosen as long as it is applied consistently. All have a zero value at their initial point and unit value at their terminal point.

The points inside the tetrahedron may be described in terms of this parametric system as

$$x = x_1 + (x_2 - x_1)u + (x_3 - x_1)v + (x_4 - x_1)w,$$
$$y = y_1 + (y_2 - y_1)u + (y_3 - y_1)v + (y_4 - y_1)w,$$

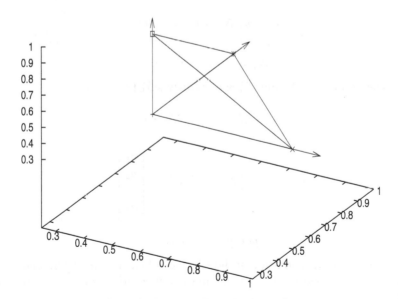

FIGURE 12.5 Tetrahedron element

and
$$z = z_1 + (z_2 - z_1)u + (z_3 - z_1)v + (z_4 - z_1)w.$$

One can verify that $(u, v, w) = (0, 0, 0)$ reduces to $(x, y, z) = (x_1, y_1, z_1)$. Similarly, $(u, v, w) = (1, 0, 0)$ reduces to $(x, y, z) = (x_2, y_2, z_2)$ and so on.

We will assume that every node point of the tetrahedron has an associated solution quantity, say, $p(x, y, z)$, throughout the three-dimensional continuum. This could be, for example, the pressure of the fluid at all spatial locations when modeling fluids.

Hence, there are four nodal solution components of the element as

$$p_e = \begin{bmatrix} p_1 \\ p_2 \\ p_3 \\ p_4 \end{bmatrix}.$$

The solution at any location inside one element is approximated with the help of the shape functions as

$$p(x, y, z) = N p_e.$$

Instead of the general procedure developed for computing the shape functions of the triangular element, we can compute the shape functions in the local, parametric coordinates very simply as

$$N_1 = u, \ N_2 = v, \ N_3 = w,$$

and
$$N_4 = 1 - u - v - w.$$

Such a selection of the N_i functions obviously satisfies

$$N_1 + N_2 + N_3 + N_4 = 1.$$

Then

$$p(x, y, z) = \begin{bmatrix} N_1 \ N_2 \ N_3 \ N_4 \end{bmatrix} \begin{bmatrix} p_1 \\ p_2 \\ p_3 \\ p_4 \end{bmatrix},$$

or
$$p(x, y, z) = N p_e.$$

Here p_i is the $p(x, y, z)$ solution of the ith node of the tetrahedron. The above equations will be used in the element matrix generation for the tetrahedron element.

12.6.2 Finite element matrix in parametric coordinates

The tetrahedron element matrix is formulated as

$$A_e = \int \int \int B^T B \, dxdydz,$$

where B is now of order 3 by 4, resulting in an element matrix of order 4 by 4. The integral can be transformed to parametric coordinates

$$A_e = \int \int \int B^T B \, det[\frac{\partial(x, y, z)}{\partial(u, v, w)}]dudvdw,$$

where

$$\frac{\partial(x, y, z)}{\partial(u, v, w)} = \begin{bmatrix} \frac{\partial x}{\partial u} & \frac{\partial x}{\partial v} & \frac{\partial x}{\partial w} \\ \frac{\partial y}{\partial u} & \frac{\partial y}{\partial v} & \frac{\partial y}{\partial w} \\ \frac{\partial z}{\partial u} & \frac{\partial z}{\partial v} & \frac{\partial z}{\partial w} \end{bmatrix} = \begin{bmatrix} x_2 - x_1 & x_3 - x_1 & x_4 - x_1 \\ y_2 - y_1 & y_3 - y_1 & y_4 - y_1 \\ z_2 - z_1 & z_3 - z_1 & z_4 - z_1 \end{bmatrix} = J$$

is the Jacobian matrix, which is constant for the 4-noded tetrahedral elements.

To compute the element matrix, the shape function derivatives for the terms of the B matrix still need to be computed. Clearly,

$$\begin{bmatrix} \frac{\partial p}{\partial u} \\ \frac{\partial p}{\partial v} \\ \frac{\partial p}{\partial w} \end{bmatrix} = J \begin{bmatrix} \frac{\partial p}{\partial x} \\ \frac{\partial p}{\partial y} \\ \frac{\partial p}{\partial z} \end{bmatrix}.$$

Then

$$\begin{bmatrix} \frac{\partial p}{\partial x} \\ \frac{\partial p}{\partial y} \\ \frac{\partial p}{\partial z} \end{bmatrix} = J^{-1} \begin{bmatrix} \frac{\partial p}{\partial u} \\ \frac{\partial p}{\partial v} \\ \frac{\partial p}{\partial w} \end{bmatrix}.$$

The terms of J^{-1} may be computed as

$$J^{-1} = \frac{adj(J)}{det(J)} = \begin{bmatrix} J_{11} & J_{12} & J_{13} \\ J_{21} & J_{22} & J_{23} \\ J_{31} & J_{32} & J_{33} \end{bmatrix}.$$

The calculation of this inverse is commonly used in industrial finite element analyses to diagnose ill-shaped elements, a common outcome of automated meshing techniques. If, in a worst-case scenario, the inverse cannot be computed, the particular element must be discarded and a new element created by modifying neighboring elements appropriately.

Furthermore, since

$$p(x, y, z) = \sum_{1}^{4} N_i p_i = u p_1 + v p_2 + w p_3 + (1 - u - v - w) p_4,$$

it follows that

$$\begin{bmatrix} \frac{\partial p}{\partial u} \\ \frac{\partial p}{\partial v} \\ \frac{\partial p}{\partial w} \end{bmatrix} = \begin{bmatrix} p_1 - p_4 \\ p_2 - p_4 \\ p_3 - p_4 \end{bmatrix}.$$

Substituting results in

$$\begin{bmatrix} \frac{\partial p}{\partial x} \\ \frac{\partial p}{\partial y} \\ \frac{\partial p}{\partial z} \end{bmatrix} = J^{-1} \begin{bmatrix} p_1 - p_4 \\ p_2 - p_4 \\ p_3 - p_4 \end{bmatrix} = \begin{bmatrix} J_{11} & J_{12} & J_{13} \\ J_{21} & J_{22} & J_{23} \\ J_{31} & J_{32} & J_{33} \end{bmatrix} \begin{bmatrix} p_1 - p_4 \\ p_2 - p_4 \\ p_3 - p_4 \end{bmatrix}.$$

Multiplication and reordering produce the terms of the B matrix as

$$B = \begin{bmatrix} J_{11} & J_{12} & J_{13} & b_{14} \\ J_{21} & J_{22} & J_{23} & b_{24} \\ J_{31} & J_{32} & J_{33} & b_{34} \end{bmatrix},$$

where

$$b_{14} = -(J_{11} + J_{12} + J_{13}),$$
$$b_{24} = -(J_{21} + J_{22} + J_{23}),$$

and

$$b_{34} = -(J_{31} + J_{32} + J_{33}).$$

With this B matrix, we finally obtain

$$\begin{bmatrix} \frac{\partial p}{\partial x} \\ \frac{\partial p}{\partial y} \\ \frac{\partial p}{\partial z} \end{bmatrix} = B p_e = B \begin{bmatrix} p_1 \\ p_2 \\ p_3 \\ p_4 \end{bmatrix},$$

as needed for the element matrix computation. The product $B^T B$ produces a 4×4 element matrix. This element matrix will contribute to four columns and rows of the assembled finite element matrix.

The element matrix is now computed as

$$A_e = \int_{u=0}^{1} \int_{v=0}^{1-u} \int_{w=0}^{1-u-v} f(u,v,w)\,dw\,dv\,du,$$

where

$$f(u,v,w) = B^T B \det(J).$$

Gaussian integration requires the additional transformation of this integral into

$$A_e = \int_{r=-1}^{1} \int_{s=-1}^{1} \int_{t=-1}^{1} \overline{f}(r,s,t)\,dt\,ds\,dr.$$

The Gaussian integration method applied to this triple integral is the Gaussian cubature introduced in Section 6.4.1.

$$A_e = \Sigma_{i=1}^{n} c_i \Sigma_{j=1}^{n} c_j \Sigma_{k=1}^{n} c_k \overline{f}(r_i, s_j, t_k).$$

In our case J and $B^T B$ are constant, hence the element matrix is simply

$$A_e = \frac{1}{6} B^T B \det(J).$$

More general finite elements require more work in executing this integration [3].

12.6.3 Local to global coordinate transformation

The element matrix for the tetrahedron element has been developed in terms of a local, oblique (u,v,w) coordinate system. Thus, before assembling any element, the element matrix must be transformed to the global coordinate system common to all the elements. The coordinates of a point in the two systems are related as

$$\begin{bmatrix} x \\ y \\ z \\ 1 \end{bmatrix} = T \begin{bmatrix} u \\ v \\ w \\ 1 \end{bmatrix}.$$

The transformation is formed as

$$T = \begin{bmatrix} u_x & v_x & w_x & x_1 \\ u_y & v_y & w_y & y_1 \\ u_z & v_z & w_z & z_1 \\ 0 & 0 & 0 & 1 \end{bmatrix},$$

where

$$\underline{u} = u_x \underline{i} + u_y \underline{j} + u_z \underline{k}$$

is the unit vector defining the u local parametric coordinate axis and $\underline{v}$ and $\underline{w}$ have similar respective roles. The point (x_1, y_1, z_1) of course defines the local system's origin.

The same transformation is applicable to the nodal degrees of freedom of any element. The global solution values are related to the local values by the same transformation matrix.

$$\begin{bmatrix} p_{e,x} \\ p_{e,y} \\ p_{e,z} \\ 1 \end{bmatrix} = T \begin{bmatrix} p_{e,u} \\ p_{e,v} \\ p_{e,w} \\ 1 \end{bmatrix}.$$

Hence, the element solutions in the two systems are related as

$$p_e^g = T p_e.$$

The p_e^g notation refers to the element solution in the global coordinate system.

Consider a solution component with the local element matrix and the local solution values. The local solution is of the form

$$A_e p_e = f_e.$$

Since the right-hand side is also given in local terms, it also needs to be transformed to global coordinates as

$$f_e^g = T f_e.$$

The solution p_e^g is also represented in global coordinates, which is the subject of interest to the engineer anyway. Introducing the global solution and right-hand side components, the solution equation reads as

$$A_e T^{-1} p_e^g = T^{-1} f_e^g.$$

Premultiplying by T results in

$$T A_e T^{-1} p_e^g = f_e^g,$$

or

$$A_e^g p_e^g = f_e^g,$$

where

$$A_e^g = T A_e T^{-1}$$

is the element matrix in global coordinates. This transformation follows the element matrix generation and precedes the assembly process.

12.7 Fluid-structure interaction application

A characteristic application of boundary value problems is the fluid-structure interaction. The wave equation, a hyperbolic partial differential equation, describes the motion of a compressible fluid as

$$\frac{1}{B}\ddot{p} - \nabla(\frac{1}{\rho}\nabla p) = 0,$$

where $p = p(x, y, z)$ is the pressure in the fluid, ρ is its density and B is the bulk modulus

$$B = c^2 \rho_0,$$

with c as the speed of sound and ρ_0 being the density of the fluid with no motion. The boundary condition imposed on the free surface of the fluid is

$$p(x, y, z) = 0,$$

meaning the pressure is ambient. The boundary condition on the fluid surface adjacent to the structure is

$$\frac{\partial p(x, y, z)}{\partial n} = -\rho \ddot{u}_n.$$

Here n is the direction vector of the outward normal of the fluid boundary and $\ddot{u}_n$ is the normal acceleration of the surrounding structure. The boundary condition is rather special, as it is not fixed. The pressure gradient is proportional to the acceleration of the surrounding (flexible) structural components.

Using the finite element method discussed in this chapter, the fluid pressure field $p(x, y, z)$ is discretized into a vector as

$$P = \sum_{i=1}^{n} p(x_i, y_i, z_i)N_i,$$

where N_i are the shape functions. The boundary conditions of the fluid-structure interface are manifested in the coupling matrix of

$$A(i, j) = \int \int N_i N_j \overline{dS},$$

where the surface integral is taken on the interface surface. The connection between the forces acting on the structure as a result of the fluid pressure is computed as

$$F = -A^T P.$$

A very important application of this process is in the computation of interior noise of automobiles or airplane fuselages. The air inside the cabins is modeled as a fluid and the pressure represents the noise.

References

[1] Galerkin, B. G.; Stäbe und Platten: Reihen in gewissen Gleichgewicht-sproblemen elastischer Stäbe und Platten, *Vestnik der Ingenieure*, Vol. 19, pp. 897–908, 1915.

[2] Keller, H. B.; *Numerical Methods for Two-Point Boundary Value Problems*, Blaisdell, 1968.

[3] Komzsik, L.; *Computational Techniques of Finite Element Analysis*, CRC Press, Boca Raton, FL, 2005.

[4] Strikwerda, J. C.; *Finite Difference Schemes and Partial Difference Equations*, Brooks/Cole, 1989.

[5] Strang, G. and Fix, G.; *An Analysis of the Finite Element Method*, Prentice-Hall, Englewood Cliffs, NJ, 1973.

[6] Zienkiewicz, O. C.; *The Finite Element Method*, McGraw-Hill, 1968.

13

Integral equations

In the simplest cases integral equations may be solved naturally by differentiation to unravel the unknown function from inside the integral. For example, consider the integral equation

$$\int_0^x u(t)dt = \frac{x}{3}u(x).$$

Differentiating both sides, we obtain

$$u(x) - \frac{1}{3}(xu'(x) + u(x)).$$

This form is now a simple homogeneous differential equation that could be rearranged by the separation of variables into the form

$$\frac{du}{u} = 2\frac{dx}{x}.$$

Integrating both sides, we obtain the solution of

$$u(x) = cx^2,$$

which is a family of second order parabolas.

Many integral equations also contain functions of two variables, like

$$\int_0^x f(x,y)dy.$$

Such integral equations may also be differentiated using Leibniz's rule as

$$\frac{d}{dx}\int_0^x f(x,y)dy = \int_0^x \frac{\partial f(x,y)}{\partial x}dy + f(x,x).$$

By evaluating the integrals in integral equations they may be turned into initial or boundary value problems, whose solutions were discussed in Chapters 11 and 12. On the contrary, it is also possible to convert initial and boundary value problems to integral equations, as shown next.

13.1   Converting initial value problems to integral equations

Let us consider the nth order ordinary differential equation initial value problem

$$\frac{d^n y}{dx^n} + c_1(x)\frac{d^{n-1}y}{dx^{n-1}} + ... + c_n(x)y = f(x).$$

We will assume continuous c_i and initial conditions

$$y(0) = d_0, y'(0) = d_1, ..., \frac{d^{n-1}y}{dx^{n-1}}(0) = d_{n-1}.$$

Denote the highest order derivative as

$$u(x) = \frac{d^n y}{dx^n}$$

and integrate from the initial condition to an unknown x as

$$\frac{d^{n-1}y}{dx^{n-1}} = \int_0^x u(t)dt + d_{n-1}.$$

This may be repeated for the subsequent terms as

$$\frac{d^{n-2}y}{dx^{n-2}} = \int_0^x (\int_0^{t_2} u(t_1)dt_1)dt_2 + d_{n-1}x + d_{n-2},$$

and so on until all derivatives are integrated and we reach

$$\frac{d^{n-1}y}{dx^{n-1}} = \int_0^x (...(\int_0^{t_2} u(t_1)dt_1)...)dt_n +$$

$$d_{n-1}\frac{x^{n-1}}{(n-1)!} + d_{n-2}\frac{x^{n-2}}{(n-2)!} + ... + d_1 x + d_0.$$

It is possible to replace the repeated integrals with a single integral as follows. Let us consider the case of a twofold integral as

$$y(x) = \int_0^x (\int_0^{t_2} u(t_1)dt_1)dt_2.$$

Applying Laplace transform on both sides results in

$$L(y(x)) = \frac{1}{s^2}L(u(t)).$$

Using the convolution theorem, this can be inverted as

$$y(x) = \int_0^x \frac{(x-t)^{2-1}}{(2-1)!}u(t)dt.$$

Generalizing this, an n-fold integral will result in

$$\int_0^x \left(\int_0^{t_n} ... \left(\int_0^{t_2} u(t_1)dt_1\right)dt_2...\right)dt_n = \int_0^x \frac{(x-t)^{n-1}}{(n-1)!}u(t)dt.$$

Substituting into the original differential equation, combining the integrals and gathering like terms results in

$$u(x) + \int_0^x \left(c_1(x) + c_2(x)(x-t) + ... + c_n(x)\frac{(x-t)^{n-1}}{(n-1)!}\right)u(t)dt =$$

$$f(x) - d_{n-1}c_1(x) - (d_{n-1}x + d_{n-2})c_2(x) - (d_{n-1}\frac{x^{n-1}}{(n-1)!} + ... + d_1x + d_0)c_n(x).$$

Summing up the terms in the integral as

$$K(x,t) - \sum_{j=1}^n c_j(x)\frac{(x-t)^{j-1}}{(j-1)!}$$

and denoting the whole right hand side as $F(x)$, we obtain the integral equation of the form

$$\int_0^x K(x,t)u(t)dt + u(x) = F(x).$$

This is a particular type of integral equation and the internal function $K(x,t)$ is called the kernel function of the integral equation.

For the commonly occurring second order initial value problem

$$y''(x) + y(x) = f(x)$$

with boundary conditions $y(0) = 0, y'(0) = 0$, applying the above process of

$$u(x) = y''(x),$$

$$y'(x) = \int_0^x u(t)dt,$$

and

$$y(x) = \int_0^x \left(\int_0^{t_2} u(t_1)dt_1\right)dt_2 = \int_0^x (x-t)u(t)dt$$

results in the integral equation

$$\int_0^x (x-t)u(t)dt + u(x) = f(x).$$

Finding the $u(x)$ solution functions of the various types of integral equations is the subject of this chapter.

13.2   Converting boundary value problems to integral equations

Let us also view how to convert boundary value problems to integral equations and consider the problem type of

$$y''(x) + p(x)y'(x) + q(x)y(x) = f(x)$$

with boundary conditions

$$y(a) = c, y(b) = d.$$

As in the last section, we introduce again $y''(x) = u(x)$ and integrate likewise

$$\int_a^x y''(t)dt = \int_a^x u(t)dt,$$

resulting in

$$y'(x) = y'(a) + \int_a^x u(t)dt.$$

Integrating this again and utilizing the first boundary condition, we get

$$y(x) = c + (x - a)y'(a) + \int_a^x (\int_a^{t_2} u(t_1)dt_1)dt_2.$$

Substituting the second boundary condition into the above equation yields

$$y(b) = d = c + (b - a)y'(a) + \int_a^x (\int_a^{t_2} u(t_1)dt_1)dt_2.$$

The $y'(a)$ value is unknown and may be obtained as

$$y'(a) = \frac{d - c}{b - a} - \frac{1}{b - a}\int_a^b (\int_a^b u(t_1)dt_1)dt_2.$$

With this the solution of the initial value problem becomes

$$y(x) = c + (x - a)(\frac{d - c}{b - a} - \frac{1}{b - a}\int_a^b (\int_a^b u(t_1)dt_1)dt_2) + \int_a^x (\int_a^{t_2} u(t_1)dt_1)dt_2.$$

Substituting the above y, y' and $y''(x)$ into the original boundary value problem, using the n-fold integral formula introduced in the last section as

$$\int_a^x (\int_a^{t_2} u(t_1)dt_1)dt_2 = \int_a^x (x - t)u(t)dt,$$

and combining the integrals, with some algebraic tedium we obtain the integral equation corresponding to the boundary value problem as

$$u(x) - \int_a^b K(x,t)u(t)dt = F(x) = f(x) - (d-c)(p(x) + xq(x)) - cq(x).$$

The kernel function

$$K(x,t) = \begin{cases} (p(x) + tq(x))(1-x); a \le t \le x \\ (p(x) + xq(x))(1-t); x \le t \le b \end{cases}$$

is of a specific symmetric form. The process described in generic terms is rather difficult but may be clarified with a simple example of

$$y''(x) + y(x) = 0,$$

with $y(0) = y_0, y(1) = y_1$ boundary conditions. For this case $p(x) = 0$ and $q(x) = 1$ and the equivalent integral equation becomes

$$u(x) - \int_0^1 K(x,t)u(t)dt = -x(y_1 - y_0) - y_0$$

with

$$K(x,t) = \begin{cases} t(1-x); 0 \le t \le x \\ x(1-t); x \le t \le 1 \end{cases}$$

as the kernel function. While for the sake of clarity the derivation above was illustrated by a specific and only quadratic form, any general and higher order boundary value problem may be converted to integral equations using similar steps repeatedly.

13.3   Classification of integral equations

The above integral equation examples were members of the several classes of integral equations occurring in practice. Integral equations may be of Fredholm or Volterra types, first or second kind, homogeneous or nonhomogeneous, and finally linear or nonlinear [2], [3].

In Fredholm type integral equations both limits of the integral are constant, for example,

$$\int_a^b K(x,y)u(y)dy + f(x) = 0.$$

This is a Fredholm integral equation of the first kind where the unknown function occurs only inside the integral. A Fredholm equation of the second kind is of the form

$$\int_a^b K(x,y)u(y)dy + f(x) = u(x),$$

where the unknown function also occurs outside the integral.

Integral equations with only one limit of the integral being constant are called Volterra types. A first kind Volterra equation is

$$\int_a^x K(x,y)u(y)dy + f(x) = 0,$$

while a second kind is

$$\int_a^x K(x,y)u(y)dy + f(x) = u(x).$$

If all terms in the equation contain the unknown function, it is called a homogeneous equation. Otherwise the integral equation is nonhomogeneous. All the above four examples were nonhomogeneous due to the presence of the so-called perturbation function $f(x)$.

On the other hand, if the unknown function occurred inside another function, like the equation

$$\int_a^x K(x,y)F(u(y))dy + f(x) = u(x)$$

or with an exponent higher than 1, the equation is nonlinear.

The kernel function $K(x,y)$ and the $f(x)$ function are assumed to be continuous in the defined domains of the particular problem. The unknown function is assumed to be continuous in the interval specified by the limits of the integral. This assumption may not always hold in practical applications.

The kernel functions $K(x,y)$ in some applications may be product kernels which are composed of products of functions of x of y only,

$$K(x,y) = g_1(x)h_1(y) + g_2(x)h_2(y) + ... + g_n(x)h_n(x),$$

and may also be symmetric when

$$K(x,y) = K(y,x).$$

The conversion of boundary value problems to integral equations shown in the last section resulted in such symmetric kernels. The specialties of the kernels may be exploited when solving a particular problem.

13.4   Fredholm solution

We start the approximate solutions with Fredholm's own solution. The idea is similar to the numerical integration concept introduced in Chapter 6, the subdivision of the interval by a finite number of intermediate points.

Let us consider a nonsymmetric Fredholm type integral equation of the second kind with a yet unknown coefficient λ as

$$\lambda \int_a^b K(x,y)u(y)dy + f(x) = u(x).$$

The unknown coefficient is called the parameter of the integral equation and its value will influence the solution.

The existence of the solution for such problems may be proven. Assuming that the integral operator of

$$\mathcal{K} = \lambda \int_a^b K(r,y)u(y)dy$$

is bounded and satisfies the Lipschitz condition (introduced in a prior chapter) of

$$||\mathcal{K}u_1 - \mathcal{K}u_2|| \le k||u_1 - u_2||$$

for $k > 0$, then the integral equation has a unique solution.

Fredholm proposed to subdivide the integral interval equidistantly with

$$\Delta = \frac{b-a}{n}$$

and compute

$$x_1 = y_1 = a,$$

$$x_2 = y_2 = a + \Delta$$

and

$$x_n = y_n = a + (n-1)\Delta.$$

The approximate solution of the integral with these becomes

$$\int_a^b K(x,y)u(y)dy \approx \Delta(K(x,y_1)u(y_1) + K(x,y_2)u(y_2) + ... + K(x,y_n)u(y_n))$$

and the integral equation translates into

$$\lambda\Delta(K(x,y_1)u(y_1) + K(x,y_2)u(y_2) + ... + K(x,y_n)u(y_n)) + f(x) \approx u(x).$$

We assume that this equation holds for all x values in the integral interval. Evaluating the known functions at the intermediate x values, we obtain a system of linear equations

$$(1 - \lambda\Delta K(x_1,y_1))u(x_1) - \lambda\Delta K(x_1,y_2)u(x_2)... - \lambda\Delta K(x_1,y_n)u(x_n) = f(x_1)$$

$$-\lambda\Delta K(x_2,y_1)u(x_1) + (1 - \lambda\Delta K(x_2,y_2))u(x_2)... - \lambda\Delta K(x_2,y_n)u(x_n) = f(x_2)$$

$$-\lambda\Delta K(x_n,y_1)u(x_1) - \lambda\Delta K(x_n,y_2)u(x_2)... + (1 - \lambda\Delta K(x_n,y_n))u(x_n) = f(x_n)$$

where the $u(x_i)$ are the values of the unknown function at the discretized locations.

Introducing the vectors

$$\underline{u} = \begin{bmatrix} u(x_1) \\ u(x_2) \\ ... \\ u(x_n) \end{bmatrix},$$

$$\underline{f} = \begin{bmatrix} f(x_1) \\ f(x_2) \\ ... \\ f(x_n) \end{bmatrix},$$

and the matrix

$$A = \begin{bmatrix} (1 - \lambda\Delta K(x_1,y_1)) & -\lambda\Delta K(x_1,y_2) & ... & \lambda\Delta K(x_1,y_n) \\ -\lambda\Delta K(x_2,y_1) & (1 - \lambda\Delta K(x_2,y_2)) & ... & \lambda\Delta K(x_2,y_n) \\ -\lambda\Delta K(x_n,y_1) & -\lambda\Delta K(x_1,y_2) & ... & (1 - \lambda\Delta K(x_n,y_n)) \end{bmatrix},$$

we write the above system of equations in the form

$$A(\lambda)\underline{u} = \underline{f}.$$

Evaluating the determinant enables us to find the value of the λ parameter which determines the existence of a solution. In the case of a homogeneous system $\underline{f} = \underline{0}$ a nontrivial solution exists only when $det(A) = 0$. In the case of a nonhomogeneous system the solution exists when $det(A) \neq 0$. In this case the solution of the integral equation is found as

$$u(x) = \lambda \int_a^b \Gamma(x, y, \lambda) f(y) dy + f(x)$$

where Γ is the so-called resolvent kernel of the form

$$\Gamma(x, y, \lambda) = \frac{\sum_{n=0}^{\infty} (-1)^n K_n(x, y) \lambda^n}{\sum_{n=0}^{\infty} (-1)^n \delta_n \lambda^n}.$$

Starting from

$$\delta_0 = 1, K_0(x, y) = K(x, y),$$

the terms may be computed recursively for $n = 1, 2, ...$ by

$$\delta_n = \frac{1}{n} \int_a^b K_{n-1}(x, y) dx,$$

and

$$K_n(x, y) = \delta_n K(x, y) - \int_a^b K(x, t) K_{n-1}(t, y) dt.$$

In practice, the summations are executed up to a finite number of terms only.

13.4.1   Computational example

We will illustrate Fredholm's method with the example equation of

$$\lambda \int_0^{\pi} \sin(x + y) u(y) dy + 1 = u(x).$$

For simplicity we divide the integral interval into three segments, e.g., $n = 3$. In practice a larger number of subdivisions is needed for reasonable approximation accuracy. In this case we compute

$$\Delta = \frac{\pi}{3}$$

$$x_1 = y_1 = 0,$$
$$x_2 = y_2 = \frac{\pi}{3},$$

and

$$x_3 = y_3 = \frac{2\pi}{3}.$$

The kernel function substitution values are

$$K(x_1, y_1) = sin(0), K(x_1, y_2) = sin(\frac{\pi}{3}), K(x_1, y_3) = sin(\frac{2\pi}{3}),$$

$$K(x_2, y_1) = sin(\frac{\pi}{3}), K(x_2, y_2) = sin(\frac{2\pi}{3}), K(x_2, y_3) = sin(\pi)$$

and

$$K(x_3, y_1) = sin(\frac{2\pi}{3}), K(x_3, y_2) = sin(\pi), K(x_3, y_3) = sin(\frac{4\pi}{3}).$$

Introducing the intermediate variable

$$a = \Delta sin(\frac{\pi}{3}) = \Delta sin(\frac{2\pi}{3}) = -\Delta sin(\frac{4\pi}{3})$$

the determinant of the system matrix becomes

$$det(A) = \begin{vmatrix} 1 & -a\lambda & -a\lambda \\ -a\lambda & 1-a\lambda & 0 \\ -a\lambda & 0 & 1+a\lambda \end{vmatrix},$$

which yields the characteristic equation

$$1 - 3a^2\lambda^2 = 0$$

whose solution is

$$\lambda = \pm\frac{2}{\pi}.$$

Since this example problem is nonhomogeneous, these are the values of the parameter for which a solution will not exist. The resolvent based solution for the example is then computed as

$$\delta_0 = 1, K_0(x, y) = sin(x + y),$$

$$\delta_1 = \int_0^\pi sin(x + x)dx = \int_0^\pi sin(2x)dx = 0,$$

$$K_1(x, y) = 0 - \int_0^\pi sin(x + t)sin(t + y)dt = -\frac{\pi}{2}cos(x - y),$$

$$\delta_2 = -\frac{1}{2}\int_0^\pi \frac{\pi}{2}cos(0)dx = -\frac{\pi^2}{4},$$

and

$$K_2(x, y) = -\frac{\pi^2}{4}sin(x + y) + \int_0^\pi sin(x + t)\frac{\pi}{2}cos(t - y)dt = 0.$$

Since K_2 became zero, the following δ_3 will also become zero, and so on, all the way to infinity. Hence in this case we are able to obtain a closed form solution. The resolvent is of the form

$$\Gamma(x, y, \lambda) = \frac{sin(x+y) + \frac{\pi}{2}cos(x-y)\lambda}{1 - \frac{\pi^2}{4}\lambda^2}$$

and the solution of the integral equation becomes

$$u(x) = \frac{\lambda}{1 - \frac{\pi^2}{4}\lambda^2} \int_0^\pi sin(x+y) + \frac{\pi}{2}cos(x-y)\lambda dy + 1.$$

Evaluating the integral, the solution simplifies into

$$u(x) = \frac{2\lambda cos(x) + \lambda^2\pi sin(x)}{1 - \frac{\pi^2}{4}\lambda^2} + 1.$$

It is easy to see that the solution does not exist at $\lambda = \frac{2}{\pi}$ as the denominator goes to zero. It is left to the reader to verify this solution by substituting into the integral equation. This example will also be used after the next section to illustrate another approximate solution technique.

13.5 Neumann approximation

This solution is also a successive approximation method after Picard, which we applied to solve initial value problems in Chapter 11. Now we will apply it to a nonhomogeneous Fredholm type equation of the second kind

$$\lambda \int_a^b K(x, y)u(y)dy + f(x) = u(x).$$

The process will start from the initial trial solution of

$$u_1(x) = f(x).$$

The successive approximations will be obtained by substituting the prior approximation into the integral equation. Hence

$$u_2(x) = \lambda \int_a^b K(x, y)f(y)dy + f(x).$$

With introducing t for clarity of the integration, the next step becomes

$$u_3(x) = \lambda \int_a^b K(x, y)f(y)dy + \lambda^2 \int_a^b \int_a^b K(x, y)K(x, t)f(t)dtdy + f(x).$$

By changing the order of the double integral, we get

$$u_3(x) = \lambda \int_a^b K(x,y)f(y)dy + \lambda^2 \int_a^b (\int_a^b K(x,t)K(t,y)dt)f(y)dy + f(x).$$

Denoting the inner integral as

$$K_2(x,y) = \int_a^b K(x,t)K(t,y)dt,$$

we obtain a recursive formulation

$$u_3(x) = \lambda \int_a^b K(x,y)f(y)dy + \lambda^2 \int_a^b K_2(x,y)f(y)dy + f(x).$$

Since $K_1(x,y) = K(x,y)$, the nth approximation is of the form

$$u_n(x) = \lambda \int_a^b K_1(x,t)f(t)dt + \lambda^2 \int_a^b K_2(x,y)f(y)dy + ... +$$

$$\lambda^n \int_a^b K_n(x,y)f(y)dy + f(x).$$

Here we generalized the expressions for the iterated kernel functions as

$$K_n(x,y) = \int_a^b K(x,t)K_{n-1}(t,y)dt.$$

The Neumann series is this sequence taken to the limit of infinity

$$u(x) = \lim_{n\to\infty} u_n(x) = \sum_{n=1}^{\infty} \lambda^n \int_a^b K_n(x,y)f(y)dy + f(x).$$

This may be written in terms of the summation as

$$u(x) = \lambda \int_a^b (K_1(x,y) + \lambda K_2(x,y) + \lambda^2 K_3(x,y) + ...)f(y)dy + f(x).$$

The Neumann series exhibits absolute convergence in the interval of

$$a \le x \le b, a \le y \le b,$$

if the parameter satisfies the condition

$$|\lambda| < (\int_a^b \int_a^b |K(x,y)|^2 dxdy)^{-1/2}.$$

If the maximum absolute value of the kernel $K(x,y)$ in the interval is M, then the integral may be replaced by

$$\int_a^b \int_a^b |K(x,y)|^2 dxdy \le M^2(b-a)^2.$$

It follows that

$$\frac{1}{M(b-a)} \le (\int_a^b \int_a^b |K(x,y)|^2 dx dy)^{-1/2}$$

and a simpler sufficient condition emerges as

$$|\lambda| < \frac{1}{M(b-a)}.$$

The Neumann series approximation is also applicable to Volterra type integral equations, such as the following nonhomogeneous Volterra equation of the second kind:

$$\lambda \int_a^x K(x,y)u(y)dy + f(x) = u(x).$$

The following discussion is for the case of $b = x$, but the approach is applicable to the opposite case. We will also assume for this case that when $y > x$ the kernel vanishes $K(x,y) = 0$. The iterated kernels following the earlier part of this section for the case when $y < x$ are

$$K_1(x,y) = K(x,y)$$

as before, but

$$K_2(x,y) = \int_y^x K(x,t)K(t,y)dt,$$

$$K_3(x,y) = \int_y^x K(x,t)K_2(t,y)dt,$$

and in general

$$K_n(x,y) = \int_y^x K(x,t)K_{n-1}(t,y)dt.$$

The Neumann series now attains the form

$$u(x) = \lambda \sum_{n=1}^{\infty} \lambda^{n-1} \int_a^x K_n(x,y)f(y)dy + f(x).$$

Exchanging the order of the summation and the integration produces a more useful form

$$u(x) = \lambda \int_a^x \sum_{n=1}^{\infty} \lambda^{n-1} K_n(x,y)f(y)dy + f(x).$$

It is proven that the Neumann series is convergent for any values of the parameter for this type of Volterra equations. We will also illustrate this method in the next section.

13.5.1 Computational example

We use the same example we used in connection with Fredholm's solution:

$$\lambda \int_0^\pi sin(x+y)u(y)dy + 1 = u(x).$$

The iterated kernel functions are computed as

$$K_1(x,y) = sin(x+y),$$

$$K_2(x,y) = \int_0^\pi sin(x+t)sin(t+y)dt = \frac{\pi}{2}(sin(x)sin(y)+cos(x)cos(y)) =$$

$$\frac{\pi}{2}cos(x-y),$$

$$K_3(x,y) = \frac{\pi}{2}\int_0^\pi (sin(x+t)cos(t-y))dt = \frac{\pi^2}{4}(sin(x)cos(y)+cos(x)sin(y)) =$$

$$\frac{\pi^2}{4}sin(x+y).$$

Observing the last two results, it is visible that the following iterations will bring the same alternative functions with different coefficients. Simply

$$K_4(x,y) = \frac{\pi^3}{2^3}cos(x-y)$$

and

$$K_5(x,y) = \frac{\pi^4}{2^4}sin(x+y).$$

Integrating these alternating functions as needed for the Neumann series solution yields

$$\int_0^\pi sin(x+y)dy = 2cos(x)$$

and

$$\int_0^\pi cos(x-y)dy = 2sin(x).$$

Substituting into the Neumann series and gathering like terms results in

$$u(x) = 2\lambda cos(x)(1 + \lambda^2(\pi/2)^2 + \lambda^4(\pi/2)^4 + ...)$$

$$+\lambda^2\pi sin(x)(1 + \lambda^2(\pi/2)^2 + \lambda^4(\pi/2)^4 + ...) + 1.$$

Computing the limit of the geometric series embedded in the parentheses, the solution of this problem becomes

$$u(x) = \frac{2\lambda cos(x) + \lambda^2\pi sin(x)}{1 - \lambda^2(\pi/2)^2} + 1.$$

This is of course identical to the solution obtained by Fredholm's method. The fact that we obtained a closed form solution here as well is due to the fact that the embedded common geometric series had a finite limit.

Since the maximum value of the trigonometric kernel is unity, the convergence criterion from the last section applies,

$$|\lambda| < \frac{1}{\pi - 0},$$

resulting in an interval of convergence as

$$\frac{-1}{\pi} < \lambda < \frac{1}{\pi}.$$

This is in synchrony with, albeit tighter than, the Fredholm solution where $\pm \frac{2}{\pi}$ were excluded.

Let us now consider the following nonhomogeneous Volterra equation of the second kind:

$$\lambda \int_0^x (x - y)u(y)dy + (x + 1) = u(x).$$

The iterated kernels are

$$K_1(x, y) = x - y,$$

$$K_2(x, y) = \int_y^x (x - t)(t - y)dt = \frac{(x - y)^3}{3!},$$

and

$$K_3(x, y) = \int_y^x (x - t)\frac{(t - y)^3}{3!}dt = \frac{(x - y)^5}{5!}.$$

The pattern of the iterated kernels is clearly visible and no more integration is needed to see the generic term. Using the last equation of the last section, the Neumann series solution becomes

$$u(x) = f(x) + \sum_{n=1}^{\infty} \lambda^n \left(\frac{x^{2n}}{(2n)!} + \frac{x^{2n+1}}{(2n + 1)!} \right).$$

Since this is proven to be convergent for any value of the parameter, we chose $\lambda = 1$, for which the solution becomes

$$u(x) = 1 + x + \frac{x^2}{2!} + \frac{x^3}{3!} + \frac{x^4}{4!} + \dots.$$

This is the familiar series expression of the exponential function, hence the solution is

$$u(x) = e^x.$$

Once again we obtained a closed form solution which in part demonstrates the strength of the method and in part is due to the example selection. In this case it is also easy to validate this solution by substituting and integrating:

$$\int_0^x (x - y)e^y dy = x(e^x - e^0) - (e^x(x - 1) - e^0(0 - 1)),$$

which yields

$$-(x + 1) + e^x + (x + 1) = e^x,$$

an identity.

13.6   Nystrom method

We recall the advantages of using strategically selected evaluation points for numerical integration in Chapter 6. Specifically the superiority of Gaussian quadrature over the Newton–Cotes class solutions of trapezoid or Simpson's method was drastic.

Nystrom proposed to utilize the same concept for the solution of integral equations. Instead of the Fredholm approach's equidistant discretization of the evaluation interval of the kernel function, Nystrom used the zeroes of the Legendre polynomial just as in the Gaussian quadrature. To travel on that avenue, we must convert the integral boundaries a, b to a unit interval $-1, 1$ by introducing

$$t = \frac{a + b - 2y}{a - b}$$

and conversely

$$y = \frac{a + b}{2} + \frac{b - a}{2}t.$$

Hence the integral occurring in a Fredholm integral equation becomes

$$\int_a^b K(x, y)u(y)dy = \frac{b - a}{2} \int_{-1}^{+1} K(x, \frac{a + b}{2} + \frac{b - a}{2}t)u(\frac{a + b}{2} + \frac{b - a}{2}t)dt.$$

Following the Gaussian quadrature procedure, we will approximate this integral as

$$\int_a^b K(x, y)u(y)dy \approx \frac{b - a}{2} \sum_{i=1}^n c_i K(x, \frac{a + b}{2} + \frac{b - a}{2}t_i)u(\frac{a + b}{2} + \frac{b - a}{2}t_i).$$

The weights c_i and parametric locations t_i are listed in Table 6.3 and will be used here as well. The actual locations are

$$x_i, y_i = \frac{a+b}{2} + \frac{b-a}{2}t_i.$$

The integral equation then translates into

$$\lambda\frac{b-a}{2}(c_1 K(x,y_1)u(y_1) + c_2 K(x,y_2)u(y_2) + \dots$$
$$+ c_n K(x,y_n)u(y_n)) + f(x) \approx u(x).$$

Introducing

$$d_i = \frac{b-a}{2}c_i$$

yields the system of equations similar to the Fredholm solution

$$(1 - \lambda d_1 K(x_1,y_1))u(x_1) - \lambda d_2 K(x_1,y_2)u(x_2)\dots - \lambda d_n K(x_1,y_n)u(x_n) = f(x_1)$$
$$-\lambda d_1 K(x_2,y_1)u(x_1) + (1 - \lambda d_2 K(x_2,y_2))u(x_2)\dots - \lambda d_n K(x_2,y_n)u(x_n) = f(x_2)$$
$$-\lambda d_1 K(x_n,y_1)u(x_1) - \lambda d_2 K(x_1,y_2)u(x_2)\dots + (1 - \lambda d_n K(x_n,y_n))u(x_n) = f(x_n)$$

where the $u(x_i)$ are the values of the unknown function at the discretized locations. The solution range of this nonhomogeneous system of equations is again dependent on the value of the parameter, similar to the Fredholm method discussed earlier.

13.6.1 Computational example

We solve the following example of a nonhomogeneous Fredholm equation of the second type

$$\lambda\int_0^1 (xy + \sqrt{xy})u(y)dy + x = u(x).$$

We will use $n = 2$, for which the parametric locations and weights from Table 6.3 are

$$t_1 = -0.5774, c_1 = 1, d_1 = 0.5,$$
$$t_2 = 0.5774, c_2 = 1, d_2 = 0.5.$$

The actual locations for our case of $a = 1, b = 0$ are

$$x_1, y_1 = \frac{1}{2} + \frac{1}{2}(-0.5774) = 0.2113,$$

$$x_2, y_2 = \frac{1}{2} + \frac{1}{2}(0.5774) = 0.7887.$$

The kernel function values at these locations are

$$K(x_1, y_1) = (0.2113 \cdot 0.2113) + \sqrt{0.2113^2} = 0.2559,$$

$$K(x_1, y_2) = K(x_2, y_1) = (0.2113 \cdot 0.7887) + \sqrt{0.2113 \cdot 0.7887} = 0.5749,$$

$$K(x_2, y_2) = 0.7887^2 + \sqrt{0.7887^2} = 1.4108.$$

The right-hand side function values are

$$f(x_1) = x_1 = 0.2113,$$

$$f(x_2) = x_2 = 0.7887.$$

We are now ready to build our system of equations

$$(1 - \lambda d_1 K(x_1, y_1))u(x_1) - \lambda d_2 K(x_1, y_2)u(x_2) =$$

$$(1 - \lambda 0.5 \cdot 0.2559)u(x_1) - \lambda 0.5 \cdot 0.5750u(x_2) \approx 0.2113$$

and

$$-\lambda d_1 K(x_2, y_1)u(x_1) + (1 - \lambda d_2 K(x_2, y_2))u(x_2) =$$

$$-\lambda 0.5 \cdot 0.5750u(x_1) + (1 - \lambda 0.5 \cdot 1.4107)u(x_2) \approx 0.7887.$$

The system matrix is of the form

$$A = \begin{bmatrix} 1 - 0.12795\lambda & -0.2875\lambda \\ -0.2875\lambda & 1 - 0.7054\lambda \end{bmatrix}.$$

The characteristic equation is

$$(1 - 0.12795\lambda)(1 - 0.7054\lambda) - 0.2875^2\lambda^2 = 0$$

or

$$0.0076\lambda^2 - 0.83335\lambda + 1 = 0,$$

whose smaller solution is 1.2134 and as such outside our integral range. Hence for any smaller value of the parameter the system is solvable. We chose for simplicity of the following $\lambda = 1$ and with this the system of

$$A\underline{u} = \underline{f}$$

yields

$$\begin{bmatrix} 0.8720 & -0.2875 \\ -0.2875 & 0.2946 \end{bmatrix} \begin{bmatrix} u(x_1) \\ u(x_2) \end{bmatrix} = \begin{bmatrix} 0.2113 \\ 0.7887 \end{bmatrix}$$

whose solutions are

$$u(x_1) = 1.659,$$

and
$$u(x_2) = 4.296.$$

The analytic solution of this equation by integration would be

$$u(x) = \frac{75}{26}x + \frac{30}{13}\sqrt{x}$$

which at the approximated locations would yield the exact solutions of

$$1.670,$$

and

$$4.325.$$

The approximated values are surprisingly close considering the very low order $(n = 2)$ of the approximation we used and attest to the strength of the approach.

13.7   Nonlinear integral equations

For simplicity we will investigate this topic with a specific nonlinear equation containing a unit kernel $K(x, y) = 1$ as

$$u(x) = f(x) + \int_a^x F(u(y))dy.$$

The solution is sought in the form of the series

$$u(x) = \sum_{n=0}^{\infty} u_n(x),$$

with the members computed by applying the successive approximation of Picard again. They are

$$u_0(x) = f(x),$$

$$u_1(x) = \int_a^x F(u_0(y))dy,$$

$$u_2(x) = \int_a^x F(u_1(y))dy,$$

and

$$u_n(x) = \int_a^x F(u_{n-1}(y))dy.$$

Then we approximate the function containing the solution in a Taylor series about $u_0(x)$ [1] as follows:

$$F(u) = F(u_0) + (u - u_0)\frac{\partial F}{\partial u}(u_0) + \frac{(u - u_0)^2}{2!}\frac{\partial^2 F}{\partial u^2}(u_0)+$$

$$\frac{(u - u_0)^3}{3!}\frac{\partial^3 F}{\partial u^3}(u_0) + \dots .$$

The use of the partial derivative reflects the fact that the function F may also contain other terms like $F(x, u)$. Since

$$u - u_0 = \sum_{n=1}^{\infty} u_n(x),$$

combining the two approximations, we obtain

$$F(u) = F(u_0) + \sum_{n=1}^{\infty} u_n(x)\frac{\partial F}{\partial u}(u_0) + \frac{1}{2!}(\sum_{n=1}^{\infty} u_n(x))^2\frac{\partial^2 F}{\partial u^2}(u_0)+$$

$$\frac{1}{3!}(\sum_{n=1}^{\infty} u_n(x))^3\frac{\partial^3 F}{\partial u^3}(u_0) + \dots .$$

Denoting the components as

$$F_0 = F(u_0),$$

$$F_1 = u_1\frac{\partial F}{\partial u}(u_0),$$

$$F_2 = u_2\frac{\partial F}{\partial u}(u_0) + \frac{1}{2}u_1^2\frac{\partial^2 F}{\partial u^2}(u_0),$$

$$F_3 = u_3\frac{\partial F}{\partial u}(u_0) + \frac{1}{2}(2u_1u_2)\frac{\partial^2 F}{\partial u^2}(u_0) + \frac{1}{6}u_1^3\frac{\partial^3}{\partial u^3}(u_0),$$

and so on, our original equation posed at the beginning of this section may be written as

$$\sum_{n=1}^{\infty} u_n(x) = f(x) + \int_a^x \sum_{n=0}^{\infty} F_n dy.$$

Expanding the summations yields the form

$$u_0(x) + u_1(x) + u_2(x) + u_3(x) + \dots =$$

$$f(x) + \int_a^x F_0(y)dy + \int_a^x F_1(y)dy + \int_a^x F_2(y)dy + \int_a^x F_3(y)dy + \dots$$

from which emerges the recursive solution as

$$u_0(x) = f(x)$$

$$u_1(x) = \int_a^x F_0(y)dy,$$

$$u_2(x) = \int_a^x F_1(y)dy,$$

$$u_3(x) = \int_a^x F_2(y)dy,$$

and so on. The recursivity is hidden in the fact that F_1 is a function of u_1, F_2 of u_2 and so on, hence the above equations must be evaluated consecutively. This series solution may be truncated at some point to provide an approximate solution and on occasion the series itself does that by vanishing terms above a certain step, hence producing a closed form solution.

13.7.1 Computational example

We illustrate the above solution by the example

$$u(x) = \frac{x^2}{2} + \int_0^x u^2(y)dy.$$

Since $F(u) = u^2$, the derivatives we need are $\frac{\partial F}{\partial u} = 2u$ and $\frac{\partial^2 F}{\partial u^2} = 2$, and all higher derivatives vanish. Following the process laid out in the last section,

$$u_0 = \frac{x^2}{2},$$

and

$$F(u_0) = u_0^2 = \frac{x^4}{4},$$

$$\frac{\partial F}{\partial u}(u_0) = 2u_0 = x^2,$$

$$\frac{\partial^2 F}{\partial u^2}(u_0) = 2.$$

Then the coefficients are

$$F_0 = \frac{x^4}{4},$$

$$F_1 = u_1 x^2,$$

$$F_2 = u_2 x^2 + u_1^2,$$

and

$$F_3 = u_3 x^2 + 2u_1 u_2.$$

Finally executing the integrals,

$$u_1(x) = \int_0^x \frac{y^4}{4} dy = \frac{x^5}{20},$$

$$u_2(x) = \int_0^x \frac{y^5}{20} y^2 dy = \frac{x^8}{160},$$

and

$$u_3(x) = \int_0^x (\frac{y^8}{160} y^2 + (\frac{y^5}{20})^2) dy = \frac{7x^{11}}{8800}.$$

Hence the approximate solution is

$$u(x) = \frac{x^2}{2} + \frac{x^5}{20} + \frac{x^8}{160} + \frac{7x^{11}}{8800} + \dots .$$

Note the continuation dots indicating that this is not a complete series. The fact that the derivatives above the second vanish does not mean that there are no more terms in the series. The subsequent terms all contain components related to the existing derivatives.

13.8   Integro-differential equations

Before we conclude this chapter with an application of integral equations, we consider a special case where both derivatives and integrals of the unknown function occur. Such equations are called integro-differential equations. For example, a Volterra type integro-differential equation is of the form

$$\int_0^x K(x,t)u(t)dt + f(x) = \frac{\partial^n u(x)}{\partial x^n}$$

with initial conditions

$$\frac{\partial^k u(x)}{\partial x^k} = b_k; 0 \le k \le n-1.$$

In order to find $u(x)$, we apply an n-fold integral operator on both sides, yielding

$$\mathcal{L}^{-1}(\int_0^x K(x,t)u(t)dt) + \sum_{k=0}^{n-1} \frac{b_k}{k!} x^k + \mathcal{L}^{-1}(f(x)) = u(x).$$

The second term on the left-hand side represents the initial conditions. We will follow the avenue used in the last section and approximate the solution in the series

$$u(x) = \sum_{n=0}^{\infty} u_n(x)$$

then substitute into the equation

$$\mathcal{L}^{-1}\left(\int_0^x K(x,t)(\sum_{n=0}^{\infty} u_n(t))dt\right) + \sum_{k=0}^{n-1} \frac{b_k}{k!}x^k + \mathcal{L}^{-1}(f(x)) = \sum_{n=0}^{\infty} u_n(x).$$

This equation may be rewritten by expanding the solution series as

$$u_0(x) + u_1(x) + u_2(x) + \ldots = \sum_{k=0}^{n-1} \frac{b_k}{k!}x^k + \mathcal{L}^{-1}(f(x)) +$$

$$\mathcal{L}^{-1}\left(\int_0^x K(x,t)u_0(t)dt\right) + \mathcal{L}^{-1}\left(\int_0^x K(x,t)u_1(t)dt\right) + \ldots \, .$$

As in the last section, the members of the series can be found recursively as

$$u_0(x) = \sum_{k=0}^{n-1} \frac{b_k}{k!}x^k + \mathcal{L}^{-1}(f(x)),$$

$$u_1(x) = \mathcal{L}^{-1}\left(\int_0^x K(x,t)u_0(t)dt\right),$$

$$u_2(x) = \mathcal{L}^{-1}\left(\int_0^x K(x,t)u_1(t)dt\right),$$

$$u_3(x) = \mathcal{L}^{-1}\left(\int_0^x K(x,t)u_2(t)dt\right),$$

and so on. As before, the series does not necessarily produce a closed form solution in the case when the coefficients vanish above a certain step because of the recursivity. On the other hand, while infinite, the series may be a representation of a known function, as we will see in the following example.

13.8.1 Computational example

Let us consider the integro-differential equation

$$\int_0^x (x - t)u(t)dt + 1 = \frac{\partial^2 u}{\partial x^2}$$

with initial conditions

$$u(0) = 1, u'(0) = 0.$$

Integrating both sides and using the initial conditions, we get

$$u(x) = 1 + \frac{x^2}{2!} + \mathcal{L}^{-1}(\int_0^x (x-t)u(t)dt).$$

We will use the twofold integral operator

$$\mathcal{L}^{-1} = \int_0^x (\int_0^{t_2} (*)dt_1)dt_2 = \int_0^x (x-t) * dt.$$

Denoting

$$1 + \frac{x^2}{2!} = u_0(x)$$

the recursive steps produce

$$u_1(x) = \mathcal{L}^{-1}(\int_0^x (x-t)u_0(t)dt) = \frac{x^4}{4!} + \frac{x^6}{6!},$$

and

$$u_2(x) = \mathcal{L}^{-1}(\int_0^x (x-t)u_1(t)dt) = \frac{x^8}{8!} + \frac{x^{10}}{10!},$$

and so on, resulting in the final solution of the form

$$u(x) = 1 + \frac{x^2}{2!} + \frac{x^4}{4!} + \frac{x^6}{6!}\cdots,$$

which is the infinite series representation of $cosh(x)$, hence the solution is

$$u(x) = cosh(x).$$

13.9   Boundary integral method application

There are many applications of integral equations in engineering sciences. Among them are equations describing various fluid wave and heat transfer phenomena as well as elasticity problems such as vibrating strings, torsional vibrations and buckling of beams [5].

The boundary integral method, a commonly used engineering analysis tool, is based on solving elasticity boundary value problems by reformulating them into integral equation form. The governing equation of three dimensional elasticity problems is Poisson's equation

$$\Delta u(x,y,z) = f(x,y,z), (x,y,z) \in \Omega$$

and the boundary of the domain Ω is denoted by Γ. The Green's function corresponding to the Laplacian operator is

$$G(P, Q) = \frac{-1}{4\pi r},$$

where $P = (x_p, y_p, z_p)$ and $Q = (x_q, y_q, z_q)$ are two points in the domain and

$$r = \sqrt{(x_p - x_q)^2 + (y_p - y_q)^2 + (z_p - z_q)^2}.$$

Applying Green's theorem [4] results in the boundary integral solution

$$u(x, y, z) = \int_\Omega G f(x, y, z) d\Omega + \int_\Gamma u \frac{\partial G}{\partial n} d\Gamma - \int_\Gamma G \frac{\partial u}{\partial n} d\Gamma.$$

The first term on the right is due to the applied load on the domain, the second term represents displacement (Dirichlet) boundary conditions and in the third term are the derivative (Neumann) boundary conditions with respect to the normal n.

The following solution is a synthesis of the approaches of the prior sections in this chapter. We discretize the boundary by a set of points $Q_j, j = 1, ...m$ with $m - 1$ segments Γ_j between the jth and $j + 1$st points. Similarly the interior is also discretized by n points $P_i, i = 1, ...n$.

With this the approximate solution of the elasticity problem is of the form

$$u(x_{p_i}, y_{p_i}, z_{p_i}) = \sum_{j=1}^m A_{i,j} u(x_{q_j}, y_{q_j}, z_{q_j}) - \sum_{j=1}^m B_{i,j} \frac{\partial u(x_{q_j}, y_{q_j}, z_{q_j})}{\partial n}$$

with coefficients

$$A_{i,j} = \int_{\Gamma_j} \frac{\partial G(P_i, Q_j)}{\partial n} d\Gamma_j,$$

and

$$B_{i,j} = \int_{\Gamma_j} G(P_i, Q_j) d\Gamma_j.$$

With the arrays of

$$\underline{u} = \begin{bmatrix} u(x_{p_1}, y_{p_1}, z_{p_1}) \\ u(x_{p_2}, y_{p_2}, z_{p_2}) \\ ... \\ u(x_{p_n}, y_{p_n}, z_{p_n}) \end{bmatrix},$$

$$\underline{v} = \begin{bmatrix} u(x_{q_1}, y_{q_1}, z_{q_1}) \\ u(x_{q_2}, y_{q_2}, z_{q_2}) \\ ... \\ u(x_{q_m}, y_{q_m}, z_{q_m}) \end{bmatrix},$$

$$t = \begin{bmatrix} \dfrac{\partial u(x_{q_1}, y_{q_1}, z_{q_1})}{\partial n} \\ \dfrac{\partial u(x_{q_2}, y_{q_2}, z_{q_2})}{\partial n} \\ \dots \\ \dfrac{\partial u(x_{q_m}, y_{q_m}, z_{q_m})}{\partial n} \end{bmatrix}$$

representing the solutions, the displacement and tangential boundary conditions, respectively, the problem solution is in a matrix form:

$$u = A\underline{v} - B\underline{t}.$$

The matrices are of n rows and m columns containing the above $A_{i,j}, B_{i,j}$ terms. Note that the kernel function in this case is singular in the domain of the integration, contrary to the assumptions we made at the beginning of this chapter. This requires special care when evaluating the integrals [6].

The advantage of this solution approach over the solution of the boundary value problem with the methods shown in the last chapter is in problems with a homogeneous interior and no load source. In such cases the solution variation in the interior is minimal and the solution is mainly dictated by the loads on, and the behavior of, the boundary.

Analyzing the acoustic behavior of closed cavities in structures is an application well suited for this method. Examples of such applications are the noise computation inside car bodies or airplane cabins.

Another prominent application area is the modeling of structural foundations for buildings, bridges and heavy machinery. The soil foundation is usually described with the boundary integral equation and the structure is with finite elements. The interaction on the boundary of the structure results in the boundary conditions presented in above equations.

References

[1] Adomian, G.; *Nonlinear Stochastic Operator Equations*, Academic Press, New York, 1986.

[2] Cochran, J. A.; *The Analysis of Linear Integral Equations*, McGraw-Hill, New York, 1972.

[3] Cordoneanu, C.; *Integral Equations and Applications*, Cambridge University Press, Cambridge, 1991.

[4] Komzsik, L.; *Applied Calculus of Variations for Engineers*, 2nd edition, Taylor & Francis, Boca Raton, FL, 2014.

[5] Mikhlin, S. G.; *Integral Equations and their Applications to Problems in Mechanics, Mathematical Physics and Technology*, Pergamon Press, Oxford, 1957.

[6] Mushkelishvili, N. I,; *Singular Integral Equations*, Noordhoff, 1958.

14

Mathematical optimization

This topic achieved immense practical importance in the past decade. It has a very humble origin in elementary calculus when seeking the extreme values of a function $f(x)$. This problem is easily solved by finding the zeroes of the derivative function

$$\frac{\partial f(x)}{\partial x} = 0.$$

Depending on the function, this solution may yield more than one extreme value; only one of those would be the global extremum, the rest would be local extrema. Furthermore, depending on the function, the problem may be classified as a convex optimization problem. A convex function's local extremum is also a global extremum.

Most practical optimization problems are also constrained and presented in terms of multivariate functions. The general convex optimization problem then is posed as

$$f(\underline{x}) = extremum,$$

under the constraints of

$$g_j(\underline{x}) \le b_j; j = 1, ..., m.$$

The independent variables are

$$\underline{x} = \begin{bmatrix} x_1 \\ x_2 \\ ... \\ x_n \end{bmatrix}.$$

In most of this chapter we will assume that f, g_j are real valued convex functions, and the constants b_j are real. The admissible domain $D \in R^n$ of this problem is always a convex set. The approximate solution of this type of problem will be the focus of this chapter.

14.1   Existence of solution

At the constrained optimization optimum location the so-called Kuhn–Tucker condition [3] applies. The condition states that at the constrained optimum the vector sum of the objective and constraint gradients must be zero with some appropriately chosen multiplying factors. Mathematically

$$\nabla f(\underline{x}^*) + \sum_{j=1}^{m} \lambda_j \nabla g_j(\underline{x}^*) = 0,$$

where the $\underline{x}^*$ is the optimal solution location in the domain. The λ_j are Lagrange multipliers.

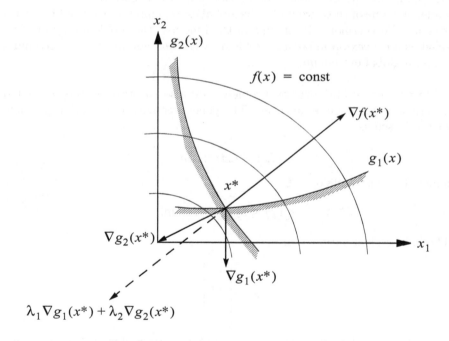

FIGURE 14.1 Optimum condition

The graphical representation of the condition is shown in Figure 14.1 with two constraint conditions represented by the shaded lines. The arcs represent the level curves of the objective function. Note that since the condition involves the gradients of both the objective and the constraint functions, they must be differentiable.

14.2 Penalty method

It is possible to transform a constrained optimization problem into an unconstrained one by introducing a new set of constraint functions defined as

$$g_j^+ = \begin{cases} g_j(\underline{x}) : g_j(\underline{x}) > 0 \\ 0 : g_j(\underline{x}) \leq 0 \end{cases}$$

with a penalty factor p and a penalty function of

$$p \sum_{j=1}^{m} (g_j^+(\underline{x}))^2$$

whose meaning is to penalize leaving the original admissible domain of the constrained problem. Then the corresponding unconstrained problem may be formulated as

$$F(\underline{x}) = f(x) + p \sum_{j=1}^{m} (g_j^+(\underline{x}))^2 = min.$$

The problem is solved by finding a sequence of p penalty factor values until they become too large. If the solution $\underline{x}$ contains the penalty factor in closed form, we may find the optimum by evaluating

$$\lim_{p \to \infty} \underline{x}.$$

The following example will illustrate this process.

14.2.1 Computational example

Let us find the solution of the optimization problem

$$f(\underline{x}) = -2x_1^2 - x_2 = min$$

with a single constraint of

$$x_1^2 + x_2^2 - 1 \leq 0.$$

Geometrically this confines our solution domain inside the unit circle of R^2.

The corresponding unconstrained optimization problem becomes

$$F(\underline{x}) = -2x_1^2 - x_2 + \begin{cases} p(x_1^2 + x_2^2 - 1)^2 : outside \\ 0 : inside. \end{cases}$$

Then the classical calculus method (for two independent variables in this case) may be used as

$$\frac{\partial F(\underline{x})}{\partial x_1} = 0$$

and

$$\frac{\partial F(\underline{x})}{\partial x_2} = 0.$$

The partial derivatives never simultaneously zero inside the circle, but outside we obtain the two equations

$$-4x_1 + p \cdot 4x_1(x_1^2 + x_2^2 - 1) = 0,$$

and

$$-1 + p \cdot 4x_2(x_1^2 + x_2^2 - 1) = 0.$$

Solving the second equation for the x_2 variable and substituting into the first, we obtain the following nonzero solution in terms of the penalty parameter:

$$x_1 = \pm(1 + \frac{1}{p} - \frac{1}{16})^{\frac{1}{2}}.$$

Backsubstituting into the second equation yields

$$x_2 = \frac{1}{4}.$$

By taking the penalty function to ∞, we obtain

$$x_1 = \pm\sqrt{1 - \frac{1}{16}}.$$

Since the objective function is quadratic in terms of the first variable, the sign of the square root is immaterial and an optimum is located at

$$\underline{x}^* = \begin{bmatrix} \sqrt{1 - \frac{1}{16}} \\ \frac{1}{4} \end{bmatrix}.$$

Hence the minimum in terms of the original function is

$$f(\underline{x}^*) = -2(1 - \frac{1}{16}) - \frac{1}{4} = -\frac{15}{8} - \frac{2}{8} = -\frac{17}{8}.$$

14.3 Quadratic optimization

In engineering practice, problems with nonlinear objective functions may also occur and quadratic objective functions are especially commonplace [6]. For example, the objective

$$f(\underline{x}) = p_1 x_1 + \ldots + p_n x_n + c_1 x_1^2 + \ldots + c_n x_n^2 = min$$

is quadratic. We gather the quadratic terms into matrix form as

$$\underline{x}^T C \underline{x}$$

where C is a diagonal matrix containing the c_i terms. Similarly we can gather the linear terms into

$$\underline{p} = \begin{bmatrix} p_1 \\ \ldots \\ p_n \end{bmatrix}.$$

With these the optimization objective is stated as

$$F(\underline{x}) = \underline{p}^T \underline{x} + \underline{x}^T C \underline{x} = min.$$

Let us present the constraints in matrix form

$$A\underline{x} \leq \underline{b}$$

where the right-hand side vector contains the constant parts (b_j) of the constraint equations and the matrix contains the coefficients of the constraint functions g_j. This assumes that the constraints are linear functions of the independent variables. The matrix A will have as many rows (m) as there are constraint equations and the number of columns will be n, the number of independent variables.

In order to have a unique solution, C must be positive definite. We will introduce a new matrix

$$G = \frac{1}{4} A C^{-1} A^T$$

that is then also positive definite and its diagonal terms are positive. Also introducing a vector

$$\underline{h} = \underline{b} + \frac{1}{2} A C^{-1} \underline{p},$$

the optimization problem may be recast into its dual form of

$$\overline{F}(\underline{u}) = \underline{h}^T \underline{u} + \underline{u}^T G \underline{u} = min,$$

subject to constraints

$$u \geq 0.$$

It can be proven that for an optimal u solution of the dual form, the optimal solution of the original problem may be found as

$$x = -\frac{1}{2}C^{-1}(A^T u + p).$$

The main advantage of this reformulation is that it lends itself to a step-wise iteration scheme that is easy to implement in MATLAB® or any other computing environment.

Starting from a zero vector $u_0 = 0$, the iterations proceed to obtain the jth term of the kth iteration by computing the intermediate value

$$u_{k,j} = \frac{-1}{G(j,j)}(\frac{1}{2}h(j) + \sum_{i=1}^{j-1}G(i,j)u_k(i) + \sum_{i=j+1}^{n}G(i,j)u_{k-1}(i)).$$

If that value is positive, then set

$$u_k(j) = u_{k,j},$$

otherwise set

$$u_k(j) = 0.$$

Even if the u sequence does not fully converge, the corresponding x provides a good approximation of the optimum of the original problem.

14.3.1   Computational example

The quadratic problem of

$$f(x) = x_1^2 + x_2^2 + x_2 = min$$

is subject to constraints

$$-x_1 - x_2 \leq -3$$
$$x_1 + x_2 \leq 5$$
$$-x_1 \leq 0$$

and

$$-x_2 \leq 0.$$

The peculiar use of the sign is to attain all constraints to be uniform inequalities. The last two are clearly stating that we are seeking positive values for the two variables. The first two imply that the sum of the two values must be greater than or equal to 3, but less than or equal to 5.

The corresponding C matrix, following the prior section, is

$$C = \begin{bmatrix} 1 & 0 \\ 0 & 1 \end{bmatrix}$$

and the $\underline{p}$ vector is

$$\underline{p} = \begin{bmatrix} 0 \\ 1 \end{bmatrix}.$$

The constraint equations are encapsulated into

$$A = \begin{bmatrix} -1 & -1 \\ 1 & 1 \\ -1 & 0 \\ 0 & -1 \end{bmatrix}$$

and

$$\underline{b} = \begin{bmatrix} -3 \\ 5 \\ 0 \\ 0 \end{bmatrix}.$$

Computing the matrices of the transformed problem results in

$$G = \frac{1}{4} \begin{bmatrix} 2 & -2 & 1 & 1 \\ -2 & 2 & -1 & -1 \\ 1 & -1 & 1 & 0 \\ 1 & -1 & 0 & 1 \end{bmatrix}$$

and

$$\underline{h} = \begin{bmatrix} -\frac{7}{2} \\ \frac{11}{2} \\ 0 \\ -\frac{1}{2} \end{bmatrix}.$$

Starting from $\underline{u}_0 = \underline{0}$ the first iteration yields all negative $u_{1,j}$ values, hence the substitution results in

$$\underline{u}_1 = \begin{bmatrix} 0 \\ 0 \\ 0 \\ 0 \end{bmatrix}.$$

Then with appropriate size unit vectors e_i we compute

$$u_{2,1} = \frac{-1}{G(1,1)}(\frac{1}{2}\underline{h}(1) + e_1^T G\underline{u}_1) = \frac{-1}{2/4}(-\frac{7}{4} + 0) = \frac{7}{2},$$

and

$$u_{2,2} = \frac{-1}{G(2,2)}(\frac{1}{2}\underline{h}(2) + e_2^T G \begin{bmatrix} \frac{7}{2} \\ 0 \\ 0 \\ 0 \end{bmatrix}) = \frac{-1}{2/4}(\frac{11}{4} - \frac{7}{4}) = -2.$$

Since this is negative and the computation of the next two terms yields

$$u_{2,3} = 0, u_{2,4} = 0,$$

therefore only the first term is nonzero in this iterative vector

$$\underline{u}_2 = \begin{bmatrix} \frac{7}{2} \\ 0 \\ 0 \\ 0 \end{bmatrix}.$$

It is left to the reader to see that the next iteration brings

$$\underline{u}_3 = \underline{u}_2 = \underline{u}^*$$

so the dual problem converged. The optimal solution of the original problem therefore becomes

$$\underline{x}^* = -\frac{1}{2}\begin{bmatrix} 1 & 0 \\ 0 & 1 \end{bmatrix}(\begin{bmatrix} -1 & 1 & -1 & 0 \\ -1 & 1 & 0 & -1 \end{bmatrix}\begin{bmatrix} \frac{7}{2} \\ 0 \\ 0 \\ 0 \end{bmatrix} + \begin{bmatrix} 0 \\ 1 \end{bmatrix}) = \begin{bmatrix} \frac{7}{4} \\ \frac{5}{4} \end{bmatrix}.$$

Clearly both variables are positive as required and their sum is less than 5, as well as equal to 3, as required by the first two constraints.

The optimal value of the objective function is

$$f(\frac{7}{4}, \frac{5}{4}) = \frac{49}{16} + \frac{25}{16} + \frac{5}{4} = \frac{94}{16}.$$

14.4   Gradient-based methods

These methods, as their name indicates, require the computation of the gradients. Their approach is in essence finding a search direction somehow related to the gradient of the objective function and proceeding in that direction until violating a constraint. At this point some evasive action takes place.

The first method of this class is the steepest descent, an idea that we met earlier in Chapter 8 in connection with nonlinear equations. We will use it now in terms of the constrained optimization problem. Using the steepest descent concept, we compute a feasible direction to progress toward the optimum of the objective function either in the direction, or the opposite, of the gradient, depending on whether the goal is a minimum or maximum.

Assuming we seek a minimum, we define a search direction as the opposite of the gradient as

$$d(\underline{x}_k) = -\nabla f(\underline{x}_k)$$

where $\underline{x}_k$ is a sequence of approximations. The direction vector is usually scaled to have its largest magnitude term equal to unity. Starting from an initial location $\underline{x}_0$ we will obtain the sequence of approximations as

$$\underline{x}_k = \underline{x}_{k-1} + \lambda_k d(\underline{x}_{k-1}),$$

for $k = 1, 2, \ldots$ until

$$\underline{d}(\underline{x}_k) = 0,$$

which means that the minimum has been reached. The value of the λ_k parameter is at first heuristically selected, for example, unit length. As a consequence of that, exactly stepping onto the location of the optimum is unlikely. The process more often steps over and away from the optimum when it is already in close proximity to it.

Another problem of this method is when a step in a certain direction becomes inadmissible because it crosses one of the constraints. Without any change of direction or step size, the process then would stop. A way to overcome this limitation is by finding an alternative feasible direction or step size.

When the direction vector crosses a constraint, the distance of the constraint from the prior location could be computed. From each constraint equation

$$g_j(\underline{x}_{k-1} + \lambda_{j,k}\underline{d}(\underline{x}_{k-1})) \leq c_j$$

we obtain a particular $\lambda_{j,k}$ value for which the jth constraint equation becomes zero. From the collection of all such values, the minimum is the largest distance not violating any of the constraints.

Using that value will now reach a particular constraint and in order to continue we devise the following strategy, called the feasible direction method. In order to present this process, let us rewrite the constraint equations, assuming they are linear, in the vector form

$$\underline{g}_1^T \underline{x} \leq b_1,$$

$$\underline{g}_2^T \underline{x} \leq b_2,$$

and

$$\underline{g}_m^T \underline{x} \leq b_m.$$

Let the vector associated with the violated constraint be the jth and create the matrix

$$R = \underline{g}_j^T.$$

A feasible continuation direction may be obtained by projecting the current direction onto the constraint being crossed. With this, the projection of the last, gradient-based direction onto the boundary results in an alternative feasible direction of

$$\underline{e}_k = (I - R^T (R R^T)^{-1} R) \underline{d}_k.$$

If the new direction vector is not a zero vector, meaning the angle between the old direction and the constraint is not 90 degrees, this produces a normal projection and the new direction is feasible (hence the name of the method).

We can find the next location as

$$\underline{x}_{k+1} = \underline{x}_k + \lambda_k \underline{e}_k,$$

and repeat the process. The length of the step λ_k may be again found by the above process.

If the new direction is a zero vector, then the current direction is perpendicular to the constraint and cannot be projected. Then the so-called modified feasible direction method [5] is often used. It is simply a heuristic deviation from the current direction vector in the range of ± 90 degrees. This is especially useful in the linear constraint case; a simple left or right turn along the violated boundary line is chosen as the new direction, as illustrated in the next example.

14.4.1   Computational example

The example problem consists of two independent variables, x_1, x_2, resulting in a two-dimensional optimization space. The objective is to find the minimum of the nonlinear function

$$f(x_1, x_2) = x_1 \cdot x_2$$

that is, the product of the two variables. Let the initial values of the variables be

$$\underline{x}_0 = \begin{bmatrix} 2 \\ 2 \end{bmatrix}.$$

The objective function at the initial location is

$$f(2,2) = 4.$$

There are three linear constraints in the problem, presented by the lines g_1, g_2, g_3 and equations

$$x_1 + x_2 \geq 2 \rightarrow g_1(x_1, x_2) = -x_1 - x_2 + 2 \leq 0,$$

$$x_2 \geq \frac{1}{2}x_1 \rightarrow g_2(x_1, x_2) = \frac{1}{2}x_1 - x_2 \leq 0,$$

and

$$x_2 \leq \frac{3}{2}x_1 \rightarrow g_3(x_1, x_2) = x_2 - \frac{3}{2}x_1 \leq 0.$$

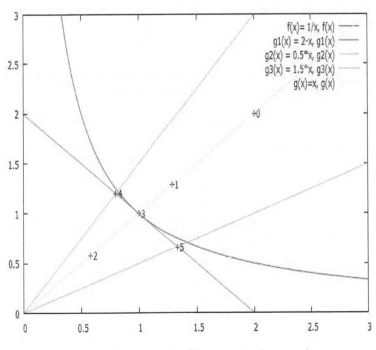

FIGURE 14.2 Gradient method example

The arrangement of the problem is shown in Figure 14.2 with the lines representing the constraints and the curve is a level curve of the objective function. The + signs and the neighboring numbers mark the sequence of locations of the optimization process.

The gradient vector of the objective function is computed by

$$\nabla f(\underline{x}) = \begin{bmatrix} \frac{\partial f}{\partial x_1} \\ \frac{\partial f}{\partial x_2} \end{bmatrix} = \begin{bmatrix} x_2 \\ x_1 \end{bmatrix}.$$

The gradient at the initial point is

$$\nabla f(2,2) = 2\underline{i} + 2\underline{j}.$$

In our example the goal is to find the minimum, hence using the gradient at the initial point we move into the opposite direction and, after scaling the gradient, we obtain the initial direction vector as

$$\underline{d}(\underline{x}_0) = -1\underline{i} - 1\underline{j}.$$

We proceed with a step in that direction with unit length, or $\lambda_0 = 1$. This will result in the new variable setting of

$$\underline{x}_1 = \begin{bmatrix} 2 - \frac{1}{\sqrt{2}} \\ 2 - \frac{1}{\sqrt{2}} \end{bmatrix}.$$

Since this step does not violate any of the constraints, we will attempt to do it again with the same direction vector and distance

$$\underline{d}(\underline{x}_1) = \underline{d}(\underline{x}_0).$$

The new setting becomes

$$\underline{x}_2 = \begin{bmatrix} 2 - \frac{2}{\sqrt{2}} \\ 2 - \frac{2}{\sqrt{2}} \end{bmatrix}.$$

This step now results in the violation of the first constraint and we find the distance to the point of violation as follows:

$$g_1(\underline{x}_1 + \lambda_1\underline{d}(\underline{x}_1)) = (-(2 - \frac{1}{\sqrt{2}} - \lambda_1) - (2 - \frac{1}{\sqrt{2}} - \lambda_1) + 2) = 0$$

from which we obtain

$$\lambda_1 = 1 - \frac{1}{\sqrt{2}}.$$

With this we obtain the new point as

$$\underline{x}_3 = \underline{x}_1 + \lambda_1 \underline{d}(\underline{x}_1) = \begin{bmatrix} 2 - \frac{1}{\sqrt{2}} \\ 2 - \frac{1}{\sqrt{2}} \end{bmatrix} + (1 - \frac{1}{\sqrt{2}}) \begin{bmatrix} -1 \\ -1 \end{bmatrix} = \begin{bmatrix} 1 \\ 1 \end{bmatrix},$$

which is of course sitting right on this constraint, as visible in Figure 14.2. The objective function at this point is

$$f(1,1) = 1$$

which is smaller than the initial value of 4, or the value at the last feasible point x_1, so we are on the right track. The gradient at this point is

$$\nabla f(1,1) = -\underline{i} - \underline{j}.$$

Since the gradient direction did not change, any move in the same direction would immediately violate the constraint on which this point is located. Therefore we have to turn away. At this point the feasible direction computation could come into consideration and we could find a new direction by the projection computation shown in the last section.

The computation for the first constraint of this problem is

$$R = \begin{bmatrix} -1 & -1 \end{bmatrix}$$

and the product becomes

$$I - R^T (RR^T)^{-1} R = \begin{bmatrix} \frac{1}{2} & -\frac{1}{2} \\ -\frac{1}{2} & \frac{1}{2} \end{bmatrix}.$$

Hence the projection

$$(I - R^T (RR^T)^{-1} R)\underline{d}_1 = \begin{bmatrix} \frac{1}{2} & -\frac{1}{2} \\ -\frac{1}{2} & \frac{1}{2} \end{bmatrix} \begin{bmatrix} -1 \\ -1 \end{bmatrix} = \begin{bmatrix} 0 \\ 0 \end{bmatrix}$$

is a zero vector because the intersection is 90 degrees, as can be seen from Figure 14.2.

Therefore we now turn to the modified feasible method; we will try 90 degree moves along the g_1 constraint until we reach any of the other two constraints. Let us first turn toward the g_3 constraint. This location is

$$\underline{x}_4 = \begin{bmatrix} \frac{4}{5} \\ \frac{6}{5} \end{bmatrix}.$$

The objective function at this point is

$$f(\frac{4}{5}, \frac{6}{5}) = \frac{24}{25}.$$

This objective has improved as it is less than the unit value obtained at the last evaluation point; however, it still may not be the optimum.

The opposite move, intersection with the g_2 constraint, results in the point

$$\underline{x}_5 = \begin{bmatrix} \frac{4}{3} \\ \frac{2}{3} \end{bmatrix}.$$

The objective function here is

$$f(\frac{4}{3}, \frac{2}{3}) = \frac{8}{9}.$$

Since this is less than the value at the other corner, we found the optimum. We verify that it satisfies the constraints as

$$\frac{4}{3} + \frac{2}{3} = 2 \rightarrow g_1,$$

$$\frac{2}{3} \geq \frac{1}{2}\frac{4}{3} = \frac{2}{3} \rightarrow g_2,$$

and

$$\frac{2}{3} \leq \frac{3}{2}\frac{4}{3} = 2 \rightarrow g_3.$$

The objective function and the first constraint were symmetric, hence the result appears contradictory. However, the other two constraints were not symmetric, imposing a bias toward a bigger value of the first variable than the second.

Hence we are at the intersection of the g_1 and g_2 constraints, and the Kuhn–Tucker condition for this point is presented as

$$\nabla f(\underline{x}_5) + \lambda_1 g_1(\underline{x}_5) + \lambda_2 g_2(\underline{x}_5) = 0.$$

The constraint gradients are

$$\nabla g_1(\underline{x}_5) = \begin{bmatrix} -1 \\ -1 \end{bmatrix}$$

and

$$\nabla g_2(\underline{x}_5) = \begin{bmatrix} \frac{1}{2} \\ -1 \end{bmatrix}.$$

The gradient of the objective function at this point is

$$\nabla f(\underline{x}_5) = \begin{bmatrix} \frac{2}{3} \\ \frac{4}{3} \end{bmatrix}.$$

Substituting, we obtain

$$\begin{bmatrix} \frac{2}{3} \\ \frac{4}{3} \end{bmatrix} + \lambda_1 \begin{bmatrix} -1 \\ -1 \end{bmatrix} + \lambda_2 \begin{bmatrix} \frac{1}{2} \\ -1 \end{bmatrix} = 0.$$

The values of $\lambda_1 = \frac{8}{9}$ and $\lambda_2 = \frac{4}{9}$ satisfy the equation, hence also the Kuhn–Tucker condition.

14.5   Global optimization

The methods presented so far were able to find a local optimum (in most cases); however, to find the global optimum of an optimization problem is elusive. The problem is even more difficult in engineering applications where a very large number (perhaps millions) of independent variables are at play. Furthermore, in such problems convexity and linearity, two conditions we relied on in the past sections, are not necessarily true globally.

The first approach to deal with this was the method of convex linearization [2]. The concept was to linearize the objective function based on the sign of its partial derivative with respect to a particular independent variable. If the sign is positive, the linearization is executed with respect to the original variable, and if negative with respect to the inverse of the variable. This resulted in a convex approximation of the original objective function with respect to all independent variables.

This concept was further developed in the method of moving asymptotes [4] with a more elaborate definition of the new variables by placing them in the interval

$$\overline{x}_i \in [\frac{1}{U_i - x_i}, \frac{1}{x_i - L_i}]$$

where U_i, L_i are upper and lower asymptotes for each independent variable.

This converts the original objective function as follows. Starting from $S^+, S^- = 0$, for each variable $i = 1, ...n$, if

$$\frac{\partial f}{\partial x_i} \geq 0,$$

accumulate

$$S^+ = S^+ + \frac{\partial f}{\partial x_i}|_{x_0}(\frac{(U_i - x_{i,0})^2}{U_i - x_i} - (U_i - x_{i,0}));$$

otherwise

$$S^- = S^- + \frac{\partial f}{\partial x_i}|_{x_0}(\frac{(x_{i,0} - L_i)^2}{x_i - L_i} - (x_{i,0} - L_i)).$$

Then we compute a new objective function as

$$\overline{f}(\underline{x}) = f(\underline{x}_0) + S^+ - S^-.$$

The asymptotes are the secret sauce in practical implementations of this method. A somewhat heuristic approach, but proven to be effective in practice, is the following. For each variable $i = 1, ..., n$ compute a starting lower asymptote as

$$L_{i,1} = x_{i,0} - \frac{1}{2}(x_{i,max} - x_{i,min}),$$

and starting upper asymptote as

$$U_{i,1} = x_{i,0} + \frac{1}{2}(x_{i,max} - x_{i,min}).$$

Then for $k = 2, 3, ...$ the iterated lower asymptote is

$$L_{i,k} = x_{i,k-1} - c_k(x_{i,k-1} - L_{i,min}),$$

and the iterated upper asymptote is

$$U_{i,k} = x_{i,k-1} + c_k(U_{i,max} - x_{i,k-1}).$$

Here the value of the constant c_k is $0.1, 1.0$ or 1.2, depending on whether the product

$$(x_{i,k} - x_{i,k-1})(x_{i,k-1} - x_{i,k-2})$$

is negative, zero or positive.

The optimization iteration is then performed as follows:

Iteration loop on $k = 0, 1, 2, ...$

1. Evaluate objective and its gradient $f(\underline{x}_k), \nabla f(\underline{x}_k)$

2. Evaluate constraints and their gradients $g_j(\underline{x}_k), \nabla g_j(\underline{x}_k), j = 1, 2, ..., m$

3. Update asymptotes $L_{i,k}, U_{i,k}; i = 1, 2, ..., n$

4. Linearize objective and constraints: $\overline{f}(\underline{x}_k), \overline{g}_j(\underline{x}_k); j = 1, 2, ...m$

5. Solve the current linearized problem of

$$\overline{f}(\underline{x}_k) = min$$

subject to constraints

$$\bar{g}_j(\underline{x}_k) \leq 0; j = 1, 2, ..., m$$

to obtain the next variable setting

$$\underline{x}_{k+1}$$

6. Increment $k = k + 1$ and repeat from 1.

The linearized problem has a unique solution that is the best approximation for the global optimum of the original problem. This method is widely used in the following engineering application.

14.6 Topology optimization

Topology optimization [1] is a very important optimization application in engineering practice with the main goal of minimizing the mass (hence weight) of a structure while maintaining or even improving its operational behavior. We will discuss this topic in a finite element framework as it is eminently suitable for such work.

The areas where elements are removed or their characteristics lowered result in a loss of mass in the structure. The mass in the finite element framework is assembled from elementary matrices and the elemental mass matrices are varied (graded) during the iterations. The total mass is therefore

$$M = \sum_{i=1}^{N} M_i = \sum_{i=1}^{N} g_i M_i^0.$$

Here $g_i \in [0, 1]$ is the grade of modification at a certain iteration and these are the independent variables of the problem, hence called design variables. The M_i^0 is the elemental mass matrix with the original, nominal characteristics.

To pose this problem to the global optimizer shown in the prior section, with the mass (weight) of the structure as objective function, we need its gradient. The mass objective derivative with respect to the design variables may be obtained as

$$\frac{\partial M}{\partial g_i} = \frac{\partial M_i}{\partial g_i} = M_i^0 = \frac{M_i}{g_i}.$$

In order to find the specific areas where material may be removed, one needs some physical constraints that can be obtained from different structural solutions. They may be deformations at certain locations in the structure, or natural frequencies. We will investigate these in the following sections.

14.6.1   Deformation constraint

Let us focus on the linear static analysis application that was introduced in Section 9.10. For simplicity of this discussion, however, we restrict the number of load vectors and therefore the set of response deformations also to only one as

$$K\underline{u} = \underline{f}.$$

As the mass matrix, the K stiffness matrix is also assembled and graded as

$$K = \sum_{i=1}^{N} K_i = \sum_{i=1}^{N} g_i K_i^0,$$

where $g_i \in [0, 1]$ is the same grade of modification at a certain iteration and the K_i^0 is the elemental stiffness matrix with the original, nominal characteristics. The gradient is again computed by exploiting the fact that each design variable is related to a single element, hence

$$\frac{\partial K}{\partial g_i} = \frac{\partial K_i}{\partial g_i} = K_i^0 = \frac{K_i}{g_i}.$$

The derivative of the static equation with respect to the ith design variable is of the form

$$\frac{\partial K}{\partial g_i}\underline{u} + K\frac{\partial \underline{u}}{\partial g_i} = \frac{\partial \underline{f}}{\partial g_i}.$$

Let us first assume that the load vector is constant and independent of the modification during the optimization, hence

$$\frac{\partial \underline{f}}{\partial g_i} = 0.$$

Premultiplying the remaining equation by the $\underline{u}$ vector, we obtain

$$\underline{u}^T K \frac{\partial \underline{u}}{\partial g_i} = -\underline{u}^T \frac{\partial K}{\partial g_i}\underline{u}.$$

Exploiting the symmetry of the stiffness matrix $(K = K^T)$ as

$$\underline{u}^T K = (K^T \underline{u})^T = (K\underline{u})^T = \underline{f}^T,$$

and substituting results in

$$\underline{f}^T \frac{\partial \underline{u}}{\partial g_i} = -\underline{u}^T \frac{\partial K}{\partial g_i} \underline{u}.$$

With the above computed gradient of the stiffness matrix the right-hand side changes into

$$\underline{f}^T \frac{\partial \underline{u}}{\partial g_i} = -\frac{1}{g_i} \underline{u}^T K_i \underline{u}.$$

The triple product is the so-called strain energy carried by the ith element,

$$\epsilon_i = \underline{u}^T K_i \underline{u},$$

where the $\underline{u}$ is the deformation vector reflecting the current stage of the modified structure. Finally we can place a constraint on the deformation of the structure at a certain location, say at the jth component of the $\underline{u}$ vector as

$$\underline{u}(j) \leq u_{max}.$$

With the corresponding load vector component, the gradient of this constraint may be written as

$$\frac{\partial \underline{u}(j)}{\partial g_i} = -\frac{1}{\underline{f}(j)g_i} \epsilon_i.$$

We now have a constraint and its gradient in addition to the objective function and its gradient, hence a proper problem for the global optimizer.

Above we assumed that the applied load remains constant while the structure is changing during the optimization process. While this is a valid assumption when the source of the load is external, there are cases when the load is changing due to the structural changes.

A frequently occurring such scenario is when a structure is optimized under the load of gravity. The acceleration of gravity at any point in a structure is the same, hence the total force of gravity (weight) is

$$\underline{f} = M \cdot a_g,$$

where the a_g is the acceleration of gravity and M is the total mass. However, since the mass of the structure is changing during the iterations, the force is not constant and the omitted gradient of the last section now becomes nonzero, specifically

$$\frac{\partial \underline{f}}{\partial g_i} = a_g \frac{\partial M}{\partial g_i} = \frac{a_g}{g_i} M_i.$$

With this the deformation constraint equation above may be appropriately modified. Similar load changing scenarios, where the measure of the load depends on the changing mass of the structure, are centrifugal or gyroscopic forces, but will not be discussed further.

14.6.2   Natural frequency constraint

The same methodology could also be applied to the free vibration (or normal modes) analysis problem shown in Section 10.8 by placing a constraint on the value of a natural frequency. The governing equation for this problem is

$$K\phi = \lambda M\phi.$$

Following the methodology of the linear static case, we differentiate with respect to the design variables as

$$\frac{\partial K}{\partial g_i}\phi + K\frac{\partial \phi}{\partial g_i} = \lambda\frac{\partial M}{\partial g_i}\phi + \frac{\partial \lambda}{\partial g_i}M\phi + \lambda M\frac{\partial \phi}{\partial g_i}.$$

We will assume that the eigenvector derivative with respect to the design variables is zero:

$$\frac{\partial \phi}{\partial g_i} = 0.$$

Left multiplying by ϕ^T and enforcing mass orthonormalization resulting in $\phi^T M\phi = 1$, the eigenvalue derivative becomes

$$\frac{\partial \lambda}{\partial g_i} = \phi^T(\frac{\partial K}{\partial g_i} - \lambda\frac{\partial M}{\partial g_i})\phi.$$

Exploiting the same finite element assembly considerations as in the last section, we obtain

$$\frac{\partial \lambda}{\partial g_i} = \phi^T\frac{K_i}{g_i}\phi - \lambda\phi^T\frac{M_i}{g_i}\phi.$$

Introducing the velocity as $\dot{\phi} = \omega\phi$ and the element kinetic energy

$$\kappa_i = \dot{\phi}^T M_i\dot{\phi},$$

with $\omega^2 = \lambda$ the gradient of the eigenvalue becomes

$$\frac{\partial \lambda}{\partial g_i} = \frac{1}{g_i}(\epsilon_i - \kappa_i).$$

This is used as the constraint gradient in connection with the above mass gradient when invoking the global optimization problem. The common industrial application scenario of this case is to minimize the weight of a structure while assuring that the first natural frequency stays above a certain minimum to avoid interference with the excitation frequencies emerging from the operational environment.

Note that both the strain and kinetic energy are standard outputs of commercial finite element software, hence their connection with a mathematical optimizer software may be done externally with a script without making any changes to either software.

14.7 Structural compliance application

While optimizing with respect to a particular deformation location or natural frequency is very valuable, in practical applications a more global consideration is often used. This is the overall strength (called compliance in industry) of the structure, defined as

$$C = \underline{f}^T \underline{u}.$$

The derivative of the compliance with respect to a design variable is

$$\frac{\partial C}{\partial g_i} = \frac{\partial \underline{f}^T}{\partial g_i} \underline{u} + \underline{f}^T \frac{\partial \underline{u}}{\partial g_i}.$$

Inserting the components found in the past two sections into the compliance derivative formula yields

$$\frac{\partial C}{\partial g_i} = \frac{\partial \underline{f}^T}{\partial g_i} \underline{u} + \underline{f}^T K^{-1} \left(\frac{\partial \underline{f}}{\partial g_i} - \frac{\partial K}{\partial g_i} \underline{u} \right).$$

Since $\underline{f}^T K^{-1} = \underline{u}^T$ and for constant forces the $\frac{\partial \underline{f}}{\partial g_i}$ terms vanish, we obtain

$$\frac{\partial C}{\partial g_i} = -\underline{u}^T \frac{\partial K}{\partial g_i} \underline{u}.$$

Due to the aforementioned assembly of the stiffness matrix, the right-hand side simplifies to

$$\frac{\partial C}{\partial g_i} = -\frac{1}{g_i} \underline{u}_j^T K_i \underline{u}_j = -\frac{1}{g_i} \epsilon_i.$$

In words, the derivative of the overall compliance of a structure with respect to a particular design variable depends on the corresponding element strain energy at a particular stage of the optimized structure.

The concept is rather simple; if an element does not carry strain, it does not contribute to the overall strength of structure and can be removed. More practically, if an element only carries a low level of strain, its structural characteristics can be lowered. Hence the ith elemental stiffness and mass matrices will change in each iteration.

FIGURE 14.3 Compliance topology optimization

The images on Figure 14.3 illustrate the structural design of a bridge using the above compliance optimization framework. The introductory shape on the upper image is a generic rectangle of which the light area is the design space allowed to be modified by the topology optimization process. The dark area is a restricted part that must be retained as it represents the roadway. The arrows are the anticipated maximum load on the structure, e.g., the weight of the maximum number of cars driving on the bridge at the same time.

The result of the topology optimization is shown in the bottom image where the amount of removed material is visibly quite significant. The shade indicates the level of compliance; the dark area is close to the limit and cannot be further reduced. Of course this shape resembles to existing bridge structures designed by experienced engineers based on their decades of knowledge. The significance of this application is that this result may be obtained automatically by a computer program hence providing valuable advice to the engineer about the physics involved.

References

[1] Bendsoe, M. P. and Sigmund, O.; *Topology Optimization, Theory, Methods and Applications*, Springer, Berlin, 2004.

[2] Fleury, C. and Braibant, V.; Structural optimization: a new dual method using mixed variables. *Int. Journal of Num. Meth, Eng.*, Vol. 23, pp. 409–428, 1986.

[3] Ponstein, J.; *Approaches to the Theory of Optimization*, Cambridge University Press, Cambridge, 1980.

[4] Svanberg, K.; The method of moving asymptotes - a new method for structural optimization. *Int. Journal of Num. Meth, Eng.*, Vol 24, pp. 359–371, 1987.

[5] Wanderplats, G. N.; An efficient feasible direction algorithm for design synthesis, *AIAA Journal*, Vol. 14, pp. 1201–1206, 1976.

[6] Wismer, D. A. and Chattergy, R.; *Introduction to Non-Linear Optimization: A Problem Solving Approach*. North-Holland, Amsterdam, 1978.

Closing remarks

This book's specific goal is to provide a working knowledge of the various approximation techniques for engineering practice. Therefore, many sections are illuminated by either a computational example or an algorithm to enhance understanding and provide a template for the reader's own application of the technique. The more advanced techniques are also illustrated by describing some of their industrial applications.

This book is designed to be self-contained. Nevertheless, each chapter also contains a reference section at the end. Some of the references are historical and mainly given to continue to record the original accomplishment. They may also be the subject of some scholarly interest. The more recent and survey references are meant to be used by the reader interested in more details about a certain method.

With an accompanying collection of MATLAB® examples, this book's original focus as a reference book may be extended and used as a textbook in classes on approximation techniques in an engineering or applied mathematics curriculum.

The additions of the second edition, the eleven new sections and the two new chapters on integral equations and mathematical optimization significantly enlarge the engineering audience reflecting the expansion of applications in the past decade.

List of Figures

List of Tables

Annotation

Notation	Meaning
$J_{n,i}(t)$	Bernstein basis polynomial
$B_{n,i}(t)$	B-spline basis polynomial
G	Gramian matrix
$G(x, y, z)$	Green's function
$H_k(x)$	Hermite base functions
$J_{i,k}(t)$	Bernstein basis
$K(x, t)$	Integral equation kernel function
$L_k(x)$	Lagrange base functions
$L(x)$	Legendre polynomials
t_i	Zeroes of Legendre polynomials
c_i	Gaussian quadrature weights
$\mathcal{L}$	Integral operator
$LS(x)$	Linear least squares approximation function
$LS_m(x)$	Polynomial least squares approximation function
$Le_k(x)$	Legendre polynomials
$S_B(t)$	Bezier spline
$S_b(t)$	B-spline
$T_k(x)$	Chebyshev polynomials
ϵ	Error of finite arithmetic, or acceptance
ϵ_s	Strain energy
e	Error of approximation
e_S	Error of Simpson's rule
e_{CS}	Error of composite Simpson's rule
e_R	Error of Romberg's method
P	Preconditioner matrix
Q	Orthogonal matrix
H	Upper Hessenberg matrix
R	Upper triangular matrix
T	Tridiagonal matrix
V	Basis for Krylov subspace
λ_i	Eigenvalue
Λ_i	Diagonal matrix of eigenvalues
ϕ_i	Eigenvector
Φ_i	Matrix of eigenvectors

K	Stiffness matrix
M	Mass matrix
B	Damping matrix
T	Tridiagonal matrix
X, Y	Lanczos vector matrices
α, β, γ	Lanczos method coefficients
B	Shape function derivatives
N_i	Shape functions
u, v, w	Parametric coordinate system
x, y, z	Cartesian coordinate system
A_e	Finite element matrix
$\underline{r}$	Location vector in Cartesian coordinates
∇	Gradient operator
Δ	Laplace operator
h, k	Finite difference step sizes
ξ, ψ	Interval derivative locations
$\underline{\dot{r}}$	First time derivative
$\underline{\ddot{r}}$	Second time derivative
$y'(x)$	Spatial coordinate derivative
L, U	Lower, upper factors or partitions
D	Diagonal factor or partition
$O(*)$	Order of *
$adj(A)$	Adjoint matrix
$det(A)$	Determinant of matrix
ODE	Ordinary differential equation
PDE	Partial differential equation
IVP	Initial value problem
BVP	Boundary value problem
J	Jacobian matrix

Index

Abel, 147
Adams methods, 240
Adams–Bashworth method, 242
Adams–Moulton methods, 243
Aitken's acceleration, 161, 162
AMG, 201
approximate derivative, 106
approximate inverse, 191
areal coordinates, 54

B-spline basis, 38
backward divided differences, 6
backward finite difference formula, 107
Bairstow's technique, 159
barycentric coordinates, 54
basis functions, 80
    linearly independent, 80
    orthogonal, 81
    orthonormal, 81
Bernstein polynomials, 29
Bessel's interpolation, 7
Bezier, 23
Bezier surface patch, 50
BFGS method, 173
biconjugate gradient method, 192
    breakdown, 193
bilinear interpolation, 285
binomial coefficients, 111
biorthogonal Lanczos method, 192
    breakdown, 193
blending function, 44
Boundary integral method, 326
boundary value problem, 271, 275
boundary value problems, 267
    of ordinary differential equations, 267

of partial differential equations, 267
bulk modulus, 301
BVP, 271, 275

Cardano's formulae, 147
centered finite difference formula, 107
    3-point, 107
    5-point, 116
    7-point, 117
Chebyshev, 79
    approximation, 90
    polynomials, 90
    quadrature, 141
closing remarks, 355
coarsening matrix, 200
column rank, 196
composite
    midpoint formula, 131
    Newton's 3/8 rule, 130
    Simpson's rule, 130
    trapezoid formula, 131
conjugate gradient techniques, 184
continuity
    first order, 33
    second order, 35
convex domain, 232
convex linearization, 345
Coons, 23
Coons surface patch, 44
Cotes, 121
coupling matrix, 301

deformation, 176
density, 301
Dirichlet boundary condition, 275, 327

Milton Keynes UK
Ingram Content Group UK Ltd.
UKHW021820071024
449327UK00021B/1372

9 780367 658076